Issues related to Augustine and just war theory are regularly debated in both popular and scholarly arenas. Rarely does an author offer a robust overview of Augustine's view while also proposing some ways to enhance the use of these ideas in a contemporary setting. Ed Herty does just that in this work. Digging deep into the context and intent of Augustine's views about just war, Herty offers a robust assessment and some very helpful suggestions for 21st century proponents of just war (both in and out of the church). Well researched and well written, this book offers a lucid and refreshing engagement with Augustine and just war.

Leo Percer, Liberty University

Jus Post Bellum and the Augustinian Worldview is clear, informative, and scholarly. The questions surrounding the future of *jus post bellum* and the long-term implications are critical to the study of lawful warfare in our time. Ed's research is profoundly helpful in contemplating the answers.

Jason Bromley, Attorney and Lay Pastor

Jus Post Bellum and the Augustinian Worldview

EDWARD HERTY

IHP ACADEMIA
West Frankfort, Illinois

Copyright © 2024 Edward A. Herty IV. All rights reserved. Except for brief quotations in critical publications or reviews, no part of this book may be reproduced in any manner without prior written permission from the publisher: admin@illativehousepress.com.

IHP ACADEMIA
An Imprint of Illative House Press
500 E. Elm St.
West Frankfort, IL 62896

IllativeHousePress.com

Paperback ISBN: 979-8-9873886-1-7
E-book ISBN: 979-8-9873886-2-4

Cover Images: Top, U.S. Army soldiers respond to an attack while conducting a patrol in a field during Operation Helmand Spider in Badula Qulp in Helmand province, Afghanistan, 15 February 2010, by Efren Lopes (Public Domain). Bottom, Genseric Sacking Rome, by Karl Briullov (Public Domain).

DEDICATION

This book is dedicated to my wonderful wife, Kathryn, and my family, especially my favorite sister, who have supported me throughout the many years of preparation for my Ph.D. I also give my profound thanks to my pastors and friends at Calvary Road Baptist Church, who have continued to provide me with their love, prayers, and support, and to my professors at Liberty University, especially Dr. Mark Allen, Dr. Edward Smither, and Dr. Leo Percer, who have served as my teachers and my mentors throughout my postgraduate education.

CONTENTS

1 Revisiting *Jus Post Bellum*   1

2 Augustine's Views on War   21

3 Just War Development and Refinement   49

4 *Jus Post Bellum* in Practice   75

5 Conclusion   123

   Bibliography   131

   Index   139

   About the Author   149

1. REVISITING *JUS POST BELLUM*

Significance of the Study

It may be impossible to fully appreciate the impact of Christianity on Western civilization. Art, music, law, ethics, and, of course, theology are but a few of the disciplines whose course changed dramatically with the advent of Christianity during the early days of the Roman Empire. Indeed, it may be extremely challenging to catalog all the ways in which the establishment of the Church has produced lasting effects that remain with the world to the present day.

Interestingly, one particular area within the study of ethics, the means by which nations and, indirectly, their respective peoples, conduct war, has a lineage that can be directly traced to Augustine. During Late Antiquity, as the Church grappled with the impact of a newly Christianized Roman government as well as a series of foreign incursions, Augustine penned his own thoughts regarding the role of the citizen and of the state in a time of national crisis.

The ideas regarding the conditions that should be present before war commences and regarding the responsibilities of a ruler and his people during a time of war remain ever-present within Western jurisprudence. As John Mattox writes, "The whole Western just-war tradition that follows from the fifth century AD on, in both its Christian and secular varieties, traces its roots not to Plato or Aristotle, nor even to earlier Church Fathers, but rather to

Augustine."[1] In fact, laws governing conduct in war today are based on Augustinian ideas regarding what should take place both before and during a conflict. As Carsten Stahn notes, "The contemporary law of armed force continues to be based on the traditional distinction between *jus ad bellum* and *jus in bello*."[2] With post-industrial technological and societal development, though, comes great change. Within the last century, a new component of just war theory has arisen in response to emerging realities.

The idea of proper conduct after the conclusion of hostilities became a serious issue after the conclusion of World War I and World War II. Even when the Treaty of Versailles was first signed at the conclusion of World War I, its signatories recognized that it contained the potential causes for a new war. Given that one conflict was a direct result of the other, the necessity of creating a lasting peace and, thus, preventing a more catastrophic future war soon became obvious. Ethical actions after a conflict had to be considered. For guidance in this area, scholars have tended to look back toward Augustinian principles.

This phenomenon creates an ethical dichotomy. Since the time of Augustine, the ideas of *jus ad bellum* and *jus in bello* have received continuous development and refinement. It is the concept of *jus post bellum* that, as a recent development, is being traced back to Augustine. If Augustine were alive today, would he agree with an idea that scholars are retroactively attributing to him? The answers to this question could be significant. First, such a study would reinvigorate the idea that the Church Fathers still speak to society in the present day. Second, this assessment has the potential to add a missing piece to Augustinian scholarship. Third, given that the concept of *jus post bellum* is still a work in progress, this study provides the opportunity to enter a Christian worldview into the debate.

Survey of Research

The various texts that will provide useful information to this study are broad in scope but tend to be disconnected. Indeed, part of the goal of this study is to provide a necessary update to the present-day understanding of Augustine's perspectives regarding the lawful conduct of war. Augustine's writings on armed conflict are numerous and have survived to the present day. Scholarly texts regarding the appropriate conduct of a nation before,

---

[1] John Mattox, *Saint Augustine and the Theory of Just War* (New York, NY: Continuum Books, 2006), 2.
[2] Carsten Stahn, "*Jus Post Bellum*: Mapping the Discipline(s)," *American University International Law Review* 23 no. 2 (2007): 326.

during, and after a conflict are also numerous and many tend to refer back to Augustinian thought. Yet, a holistic comparison of all elements of just war theory with Augustine's writings remains elusive.

Given the amount of scholarly thought in the field of just war theory, it is important that this study focuses on works that demonstrate the development of two pillars of just war theory to the twentieth century, the emergence of the new doctrine of *jus post bellum*, and the application of *jus post bellum* in practice. This survey of research will clarify the apparent disconnect between the study of lawful warfare from Augustine until the early twentieth century and the study of *jus post bellum* after World War II. It will be divided into four sections: Augustinian views on war, formulation of *jus post bellum*, the status of *jus post bellum* scholarship, and *jus post bellum* in practice. Concurrently, it will note the gap in research that this work intends to fill.

Augustinian Views on War

Augustine of Hippo – *The City of God, A Treatise Concerning the Correction of the Donatists*, and *Reply to Faustus the Manichaean*

These sources are critical to the study because they contain Augustine's exact thoughts regarding the conduct of a just war. Just war theory uses terms that, while traced to Augustine, were not coined by Augustine himself. Therefore, these primary sources are crucial in examining what Augustine said regarding *jus ad bellum* and *jus in bello* as well as what he would have to say about *jus post bellum*. In *The City of God*, Augustine redefines the roles of the sovereign and of the Christian within the context of a Christian state. In *Reply to Faustus*, Augustine defends the idea that Christians can be involved in conflict as long as they are justly led and have a rightful intention.

Of note, *The Treatise Concerning the Correction of the Donatists* and some of Augustine's letters will also be referenced during this work. Within his letters are some of his most pertinent views regarding the need to create a lasting peace and even, perhaps, some of the responsibilities of the victorious nation. As for the issue of Augustine and the state-led persecution of the Donatists, this study will briefly examine some of his thoughts on the right of the state to intervene in a problem that deeply affected his own ministry, such as the state-sanctioned seizure of Donatist property, but it will not delve too deeply into the issue as much of it lies outside the boundaries of this research.

Edward Herty

John Mark Mattox – *Saint Augustine and the Theory of Just War*

    This work is important to this study because it provides a great deal of historical background, helping explain why Augustine found a need to express his views on the idea of a just war. Mattox reviews and critiques Augustine from theological and philosophical points of view and even notes how Augustine's views may have changed over the course of his own lifetime as he experienced the collapse of the Western Roman Empire firsthand. The author provides a significant level of interaction with the thoughts presented in *The City of God* and *Letter to Faustus*. However, Mattox only speaks about *jus ad bellum* and *jus in bello*, leaving the topic of this text open for further research.

Peter Heather – *The Fall of the Roman Empire*

    This study examines war and Roman society during the time of Augustine from a historical, but not necessarily Christian, point of view. As such, it provides necessary insight into the state of Roman society during the era of Augustine, as well as an account of the rise and decline of the Roman army. The author's assertions can be compared to those made by Augustine in *The City of God*, when he claims that corruption brought about Rome's civic and, thus, military failures.[3] For this study, the passages that unpack the sack of Rome in AD 410 and the invasion of North Africa by the Vandals will prove especially useful.

Charles J. Scalise – "Exegetical Warrants for Religious Persecution: Augustine vs. the Donatists"

    This work explains a challenging portion of Augustine's historical context. In a sense, it represents religious persecution as a deviation from a present-day understanding of just war theory that must be addressed as part of Augustine's overall view of conflict. Also, in a sense, it indicates that Augustine held views regarding the separation of church and state that informed his idea of just war. It is apparent that Augustine saw the Church and the state as one unified body, as discussed when he noted that all property belonging to Christians truly belongs to the Church.[4] As noted, this study will not examine Augustine's views regarding internal conflict too deeply. However, his views on property rights are relevant in a discussion about the

---

[3] Augustine of Hippo, *The City of God, Books I-VII*, trans. Demetrius B. Zema and Gerald G. Walsh (Washington, D.C.: The Catholic University of America Press, 2008), 129-130.

[4] Charles J. Scalise, "Exegetical Warrants for Religious Persecution: Augustine vs. the Donatists," *Review and Expositor* 93 (1996): 499.

degree to which a member of a belligerent government can be held personally accountable after the conclusion of hostilities.

Daniel H. Weiss - "Christians as Levites: Rethinking Early Christian Attitudes toward War and Bloodshed via Origen, Tertullian, and Augustine"

This work discusses an interesting dichotomy in Augustinian thought regarding the role of the Christian before and after the Christianization of the Roman Empire. Weiss provides the theological background behind Augustine's assertion that Christians had a duty to nonviolence while under a pagan government before they had a duty to support the Roman army under the reign of Constantine.

Andrej Zwitter and Michael Hoelzl - "Augustine on War and Peace"

This journal article provides a necessary discussion regarding the Augustinian concept that drives most *jus post bellum* scholarly thought. The authors define post-war peace as a state of mutual harmony between themselves and with God. In their view, Augustine looked forward to a lasting, universal peace.[5] This article goes through great lengths to define the Augustinian ideal of peace in both interpersonal and theological terms. The missing component of this scholarship, though, is, in practical terms, how this peace may be achieved.

Philip Wynn - *Augustine on War and Military Service*

The author asks a similar research question to the one posed by this study. However, he attempts to tie all of just war development since the twentieth century to Augustinian thought, whereas this work will focus specifically on *jus post bellum*. Of additional note, given the date of Wynn's research, this study will be able to include a greater degree of twenty-first-century *jus post bellum* context. Specifically, with the very recent conclusion of American involvement in Afghanistan, *jus post bellum* implications of the conflict are open to reassessment.

---

[5] Andrej Zwitter and Michael Hoelzl, "Augustine on War and Peace, "*Peace Review* 26, no. 3 (August 2014): 323.

Edward Herty

Formulation of Jus Post Bellum

Daniel R. Brunstetter and Cian O'Driscoll - *Just War Thinkers: From Cicero to the 21st Century*

  This text will clarify an important issue that will be presented within this study. Namely, Augustine only provided the base ideas of just war theory. As this work discusses, the actual refinement of the theory can be attributed to thinkers such as Aquinas and the medieval thinkers who came after him. Therefore, Brunstetter's work provides necessary information as this work progresses from the Augustinian context to the present day.

  Also of note, the authors make an effort to discuss the impact of war in the present day when waged upon people of cultures that do not share the same Western Just War tradition. For example, the multi-year conflicts in the Middle East have many parallels to the conflicts between the "civilized" and "non-civilized" groups during Late Antiquity and the Middle Ages. The idea is that what appears to be just in a Western context may be seen differently through the eyes of others. This is an important point as this research will show that the same *jus post bellum* principles, when placed into practice, create very different outcomes, depending on the culture to which they are applied.

Nico Vorster - "Just War and Virtue: Revisiting Augustine and Thomas Aquinas."

  This journal article is a commentary on just war theory that attempts to translate Augustinian thought through Aquinas and into a present-day context. Of note, the author does not make the connection between Augustine and *jus post bellum* that will be found within this study. Instead, he focuses on the *jus ad bellum* and *jus in bello* aspects of just war theory. Since he speaks about present-day conflict, though, without reference to *jus post bellum* principles, it is apparent that a connection needs to be made between these parallel schools of thought.

  Of interesting note, though, the article does provide a nuanced view of the American conflicts in Iraq and Afghanistan. Essentially, he finds that, while the wars themselves may have been justified, incorrect motivations had poisoned both enterprises. Given that this study partially seeks to discern why *jus post bellum* efforts in these two areas were less than successful, Vorster's work is relevant to this study's conclusions. It is possible that the efforts were hindered by a lack of moral clarity on the part of American leadership, but it is also possible that the citizens of Iraq and Afghanistan must share some level of responsibility for the failure of reconstruction efforts. The research that will be conducted within this work will make this issue clearer as it describes the differences between the successes after World War II and failures of *jus post bellum* application in later years.

Gregory M. Reichberg - *Thomas Aquinas on War and Peace*

Reichberg's work has tangential importance to this study. He goes through great lengths to trace many of the principles espoused by Aquinas, as derived from Augustinian thought to the present day. Notably, he describes important twenty-first-century issues of the ethics of war. He goes to great lengths to trace the history of wars of self-defense and pre-emptive wars through medieval thinkers and to the present day. He also discusses issues such as whether it is permissible to remove a tyrant, the legal structures surrounding the topic of punishment, and the need to maintain a sense of legitimate authority.

All of these developments are accomplished without referring to the topic of *jus post bellum*, as defined in this work. The author does touch upon some of the criteria, such as the need to establish justice during a war and the possible need to reform a foreign government. However, a formal discussion of the various requirements of a *jus post bellum* environment, such as the need to rebuild a nation's economic structures, is noticeably absent despite the length and comprehensiveness of the work. This issue makes the need for *jus post bellum* research, as related to Augustine, more apparent.

James Turner Johnson - "Aquinas and Luther on War and Peace: Sovereign Authority and the Use of Armed Force"

The purpose of Johnson's work is to trace the development of the concept of sovereign authority from the time of Aquinas to the present day. As a component of this work that seeks to make the connections between just war theory elements that have been processed over time and the element of *jus post bellum* that has received only recent attention, this idea of sovereignty is important. In short, the concept of sovereign authority can be traced back to Augustine's requirement that only a righteous, legitimate ruler has the authority to order a war's commencement. The addition of Lutheran thought is especially interesting, given Luther's cultural context.

Of additional importance, the very idea of rulership has changed greatly since the time of Augustine. Governments since that time, as a whole, have tended to devolve authority to the people. In Augustine's time, Rome was led by an emperor and was long past its days as a republic. In Luther's time, Germany was a collection of states, led by princes who were nominally obedient to an emperor. In later years, republican forms of government emerged. Yet, the idea of sovereign authority remained. Johnson's work helps trace the evolution of this component of Augustinian thought through these various civic evolutions. Still, as with other works described in this literature review, discussion of *jus post bellum* ideals is noticeably absent.

Status of Jus Post Bellum Scholarship

Gary J. Bass - "*Jus Post Bellum*"

This article tends to focus on the duties and responsibilities of reconstruction. Specifically, Bass argues that reconstruction should take place in formerly genocidal states but not necessarily in others.[6] The issue is that state sovereignty must be preserved wherever possible. Of course, this is a complicated argument as each variant of injustice has a different dynamic. It is possible for a state to be led by oppressors who, while not engaged in genocidal acts, have lost the legitimacy of rulership. Looking through the Augustinian Just War lens of states being required to perform those actions that reduce suffering, Bass finds that most other efforts tend to produce more problems in the long term.[7] Though his thesis is not entirely convincing, as related research it is quite useful to this study.

Bass' argument is also useful when it is applied to the three case studies that will be presented later in this work. For Japan and Germany, for instance, it is apparent that reconstruction efforts were successful. For Iraq and Afghanistan, the results are much more complicated. It is possible that the levels of corruption in these states made true reconstruction impossible. It is also possible that the United States failed to adequately provide what was needed. Cultural issues may also have been relevant as European, Asian, and Middle Eastern cultures are vastly different. In answering the question regarding whether or not Augustine would approve of reconstruction, these issues must be considered

Louis V. Iasiello - "*Jus Post Bellum*"

This article studies some of the moral duties of a nation after a conflict in light of Augustinian thought. The author examines the need for

---

[6] Gary J. Bass, "Jus Post Bellum," *Philosophy and Public Affairs* 32, no. 4 (2004): 387.
[7] Ibid., 393.

war crime trials and deep national ownership of the defeated enemy's reconstruction process until well after the conclusion of hostilities. Importantly, this work implies that complying with the need for a lasting peace comes with great cost on the part of the victor and that this cost must be considered beforehand.

This, too, is a complicated argument. In all of the case studies that will be presented within this study, it will be apparent that cost considerations were not made before the commencement of the reconstruction processes. It may seem to be a natural consideration, but these types of estimates are difficult to achieve before the full extent of any postwar damage is known. Furthermore, as was the case in Afghanistan, reconstruction requirements changed over time. It is possible that under Augustinian thought, practical *jus post bellum* estimates should be made before the commencement of hostilities. Therefore, this work will be useful to the development of this work's arguments.

Albert W. Klein - "Attaining the Post Conflict Peace Using the *Jus Post Bellum* Concept"

This journal article is useful to this research because it performs two critical actions. First, it attempts to provide an overview of the field of scholarly opinions regarding the definition of *jus post bellum*. Klein consolidates the various fields of thinking into a concise format that lists the positions of over a dozen scholars on six areas of *jus post bellum* doctrine.[8] These areas have been consolidated into the four elements of *jus post bellum* that are being reviewed in this study.

Second, Klein's comprehensive view of just war theory scholarship indicates that a full consensus on this topic has not yet been reached. For the purposes of this study, the gap leaves room for a reassessment of Augustinian thought. If differing scholars have separate opinions regarding the need to rebuild, war crime trials, and humanitarian assistance, then there is room for future Augustinian scholars to enter their own research into the field.

Carsten Stahn, Jennifer Easterday, and Jens Iverson – *Jus Post Bellum: Mapping the Normative Foundations*

---

[8] Albert W. Klein, "Attaining the Post Conflict Peace Using the *Jus Post Bellum* Concept," *Religions* 11, no. 4 (2020): 184.

This book is important to this study because it examines the current state of international law as related to *jus post bellum*. This overview includes the duty of one nation to occupy the other, the requirements for reconstruction, the need to establish an updated system of social justice, and the possibility to establish a more righteous regime as part of an exit strategy.

It is important to include some elements of legal research as a component of this work. Just war theory has, over time, become codified into law. Principles such as reducing suffering and selecting the correct reasons to enter into a conflict are more than moral considerations. There are also legal ramifications as international courts have the ability to hold persons accountable after a conflict. If the principles of *jus post bellum* already have legal standing, then the need to trace these ideals back to Augustine, as was accomplished with the other components of just war theory, is even more apparent.

Jus Post Bellum in Practice

Eric D. Patterson - *Ending Wars Well: Order, Justice, and Conciliation in Contemporary Post-Conflict*

This study will be useful to this work because it focuses on the practical aspects of *jus post bellum* as applied recently in both Iraq and Afghanistan. As such, it demonstrates secular, non-scholarly thought on how wars should be ended. The author's conclusion is especially poignant as he argues that, from a historical perspective, *jus post bellum* activities have been most successful when they attempt to create conciliation between all parties.[9] This research may help partially explain why some *jus post bellum* activities have been more successful than others. Of note, though, a gap in research does exist as this book was written well before the conclusion of Western involvement in Afghanistan. Therefore, more recent journal articles may be used to supplement this work.

Michael Holm – *The Marshall Plan*

As what will become one of the two most successful *jus post bellum* outcomes, a study of the Marshall Plan will be an important feature of this study. As a historical study, this work provides the details behind the success

---

[9] Eric D. Patterson, *Ending Wars Well: Order, Justice, and Conciliation in Contemporary Post-Conflict* (New Haven, CT: Yale University Press, 2012), 162.

of this plan. Furthermore, Holm's study explains how the Marshall Plan influenced later *jus post bellum* decisions.

Of additional note, one of the challenges of this *jus post bellum* study lies in its ability to discern pure motives in *jus post bellum* activities from political pragmatism. In the post-World War II environment, specifically, it is difficult to separate genuine humanitarian aid from the political calculus that was designed to slow the spread of Soviet influence. As the work concludes, the author is careful to make this distinction.

Mark David Hall and J. Daryl Charles – *America and the Just War Tradition*

Continuing the post-World War II research, Hall and Charles provide an interesting perspective regarding American reconstruction efforts in Japan. Important distinctions are made regarding Japanese and American conduct during the war, which will prove to be an interesting component of this study as these actions will be placed in contrast with the rebuilding efforts after the war.

The authors also review *jus post bellum* activities in Iraq. Notably, they find that much of the failure of *jus post bellum* activities in that region could be attributed to a lack of pre-war planning.[10] The idea is that the Bush administration placed a great deal of emphasis on winning the war in Iraq but did not adequately plan for post-war activities. In their review of the war in Afghanistan, though, the authors trace the failure of *jus post bellum* activities back to an original failure to apply other aspects of just war theory, such as *jus ad bellum*, before the war began.[11] This adds an interesting and holistic perspective to the *jus post bellum* research that will be conducted in this study.

Kevin Jon Heller – *The Nuremberg Military Tribunals and the Origins of International Criminal Law*

War crime trials form a significant portion of recent *jus post bellum* activities. The Nuremberg Tribunals were, perhaps, the first and most historically significant applications of the principle of providing justice after a war through personal accountability.

---

[10] Mark David Hall and J. Darryl Charles, eds., *America and the Just War Tradition: A History of U.S. Conflicts* (Notre Dame, Indiana: University of Notre Dame Press, 2019), 265-266.
[11] Ibid., 282.

This work is significant because it provides insight into the trials to a very high level of detail. The author examines the origins of the trials, the crimes that were charged, the defenses, and the final sentences. Importantly, the author reviews the aftermath of the trials and their impact on decisions to make similar *jus post bellum* decisions in the future.

Need for the Current Study

Given the apparent gap in research, this project has the potential to make a notable contribution to Christian scholarship as related to the subfields of ethics and Augustinian history. First, this research will seek to add depth to a field of current legal research that loosely attributes its lineage to Augustine. While the concept of *jus post bellum* has an important connection to an Augustinian ideal, it did not receive centuries of refinement through a Christian lens unlike other concepts, such as *jus in bello* and *jus ad bellum*, which were processed through sacred and secular history. By asking whether or not *jus post bellum*, as it is understood today, can truly be attributed to Augustine, this study seeks to establish at least some of that necessary Christian refinement.

Second, since the idea of *jus post bellum* is not fully formed, there is an opportunity to add a Christian worldview to the conversation. At present, the ideas about proper conduct after the conclusion of hostilities range from the signing of a just peace to a complete overhaul of the defeated nation's economic and governmental structures. Additional elements such as war crime trials for the purpose of accountability and the possible need to provide extended humanitarian aid only add to this complexity. By developing a present-day Christian understanding of exactly what Augustine meant by the necessity of establishing a lasting peace, the Church may be able to influence the life-changing decisions that are often made by governmental entities.

Third, in reference to Augustinian scholarship, it is a given that Augustine's ideas about war were developed within a certain context. He was attempting to define the role of the Christian citizen under a Christian government. At the time, this was relatively unknown territory. Though Augustine provided valid overarching ideals, these concepts were formulated in the midst of an epochal change during Late Antiquity in the Western Roman Empire. A newly Christian empire was collapsing, and, at all corners of the empire, pagan influence was resurgent.

Given the vast changes in technology, politics, communications, and information since that time, it may be useful to compare Postmodern Era ideas regarding actions to be taken at the conclusion of hostilities to Augustine's own historical context to learn whether or not he would have agreed with them. Furthermore, given that the idea of Christian rulership in the West is fading with the present shift from modernity to late modernity and from Christendom to post-Christianity, it may be possible to draw some useful parallels. While this work may not find a simple "yes" or "no" answer to some of these questions, a nuanced conclusion followed by practical applications would certainly be appropriate.

Research Questions

Given the likelihood of a highly nuanced conclusion to the study, a wide variety of research questions will be addressed. Each question, though, will have an appropriate place within the thesis and within the research methodology. These questions will be roughly divided into three categories.

The first set of research questions addresses Augustine's own words and their historical context. Regarding the requirements that must be met before a state engages in war, Augustine implies that several conditions must be met. For example, a lawful sovereign must be the head of the state and the war must be conducted for the right reasons.[12] These are complex ideas that must be unpacked. In Augustine's mind, what is a lawful sovereign? What is a rightful intention? Regarding the requirements that must be met during the conduct of war, Augustine explains that unnecessary suffering must be avoided.[13] Given the violent nature of war and the difficulties in restraining individual combatants in his era, what would Augustine describe as excessive suffering? Finally, regarding conduct after a war has concluded, what exactly does Augustine have to say? This study is primarily concerned with this last question, but the other questions are relevant given that conduct after war is seen as a component of the larger just war theory construct.

The next set of questions seeks to apply Augustine's principles to the present day, though they tend to have more of a focus on conduct after a war has concluded. As an overarching question, what responsibilities does a victorious nation have towards its defeated enemy, and should national interests be a part of this calculus? This leads to several sub-questions that will be explored as a part of the research methodology. Should individual leaders be held accountable through war crime trials? Does the victorious nation have an obligation to care for refugees? Should the government of the belligerent nation be replaced with a more equitable, righteous body of

---

[12] Augustine, *The City of God: Books XVII-XXII*, 206-207.
[13] Augustine, *The City of God: Books I-VII*, 19.

leaders? Finally, should the victorious nation rebuild the economy and infrastructure of the defeated nation? These questions all seek to unpack what Augustine meant when he advocated for a "universal peace."[14]

Key Definitions/Concepts

Jus Ad Bellum

Jus ad bellum is a Latin term that will be used frequently during this study. Literally, the meaning of the term is "the law to war." The term refers to the actions and attitudes that must exist before a state and its sovereign are allowed to engage in what Augustine would consider to be a just war. The term originates with the ideas of Thomas Aquinas, who sought to process and consolidate Augustine's various writings on this topic.[15] He created three criteria that must be present before the decision to enter a war could be considered righteous. Other criteria have been added since that time but, for the purposes of this definition, only these three criteria will be discussed.

First, a legitimate sovereign must order the state to enter into a war.[16] This idea has several implications. A legitimate sovereign can be defined as a head of state that has authority from God to protect the state from its enemies.[17] Naturally, not every nominally Christian leader could be defined as an ardent believer. Even Emperor Constantine's faith was in doubt. The overarching idea is that the person who is ordering a state to go to war is the person who is generally recognized as having secular authority over the state. Another implication of this principle is that Augustine and, by extension, Aquinas are not referring to individual-level conflict inside the state. As Nico Vorster notes, those types of conflict are resolved by the proper authorities within the state.[18] As with instances of private internal violence in the present day, these types of conflicts are handled by internal law enforcement.

Second, the war must be waged for a just cause, meaning that the rightful sovereign is attempting to alleviate some injustice.[19] The idea is that a rightful sovereign, seeing that evil is taking place in a neighboring country, steps in to correct that evil and, thus, restore peace. In practical terms, this

---

[14] Zwitter and Hoelzl, "Augustine," 321.
[15] Nico Vorster, "Just War and Virtue: Revisiting Augustine and Thomas Aquinas," *South African Journal of Philosophy* 34, no. 1 (March 2015): 61.
[16] Ibid.
[17] Ibid.
[18] Ibid.
[19] Ibid.

concept means that a war cannot be waged solely for national interest but, rather, to promote a universal good. To put this another way, the violence of war is necessary in order to suppress an ever-higher level of injustice.

Naturally, this is a complicated criterion. Where does a righteous cause end and national interest begin? World War II, for example, could be considered a standard case of fighting a foreign enemy in order to protect other nations from oppression. Still, it is evident from the results of the war that national interest was a part of the calculus. To sum up this concept of the just cause, a war cannot be justified if the sovereign engages an enemy solely to increase his nation's standing among others.

Third, the decision to enter into a war must be made with a righteous intention in mind.[20] This idea is somewhat different from the second criterion. In Aquinas' mind, it is unjust for a state to plan to seize another state's resources even if the war is being waged for the correct reasons. It would also be unjust to make plans that include causing a disproportional level of suffering. To use a historical example, it may have been just for the Soviet Union to choose to repel foreign invaders in 1941. However, any intention that included seizing foreign territory after the war or inflicting unnecessary suffering on the enemy's civilian population would have made the entire effort unjust, according to just war theory.

Again, this is a complicated criterion. Even under a just war, foreign territory may be seized, or foreign goods may be taken. After all, a case can be made that forcing the unjust party to pay reparations could be justified. Holding foreign territory indefinitely in order to maintain a permanent peace could also be justified. The true discriminating factor lies in the sovereign's overall state of mind. If the sovereign reluctantly seeks to end suffering at no material gain to himself or to his nation, then he could be considered to have a righteous intention.

Jus In Bello

Jus in bello is a Latin term that means, literally, "justice in war." The term refers to those criteria, found in Augustine's writings, which dictate how an army should fight a just war. Naturally, it is important that a nation go to war for the right reasons. However, once hostilities commence, the armies themselves require strict guidance in order to ensure that their conduct remains lawful, at least in Augustine's eyes. The overall idea is that, since war

---

[20] Ibid., 61-62.

is, by its nature, a violent enterprise that is undertaken only with great reluctance, every effort should be made to prevent more suffering than is absolutely necessary.[21] To that end, the idea of *jus in bello* can be subdivided into three principles.

First, Augustine advocated for what is known in the present day as proportionality, meaning that the level of violence utilized by the righteous nation should not be more than is absolutely necessary in order to restore peace.[22] Though this appears to be a simple concept, it has a very broad range of applications. One implication is that, under just war theory, a nation should not cause more fatalities than is necessary. To use an example, if it is apparent that an enemy army is hopelessly outnumbered and lacks the will to fight, then peace terms should be offered. As a corollary to this idea, unnecessary wounding should also be avoided.

A second implication is that an enemy should be allowed to surrender. Once enemy combatants have made it known that they are willing to quit fighting, then every effort should be made to provide aid and comfort to them. This idea extends to the proper treatment of prisoners of war. A third implication is that individual combatants are required to behave in a professional manner. They are to avoid vengeful behavior and any type of plunder-seeking behavior.

Augustine also advocated for what is known in the present day as discrimination, meaning that soldiers, as representatives of the state, have the sole legal responsibility to administer violence.[23] To an extent, this means that soldiers should refrain from using violence against civilians and protected groups, such as medical personnel and clergy. Notable exceptions exist, though, as a civilian who takes up arms will suddenly become a combatant.

A second implication of the idea of discrimination is that soldiers have a duty to obey the orders of the sovereign. This means that the state should guide the ethical decisions of its soldiers. It also means that, as agents of the state, soldiers do not have the right to refuse to commit lawfully ordered violence.

Finally, Augustine argued that lawful conduct in war means that an army should act in good faith, meaning that deceptive military tactics are permissible, but leaders have an obligation to obey the standard rules of

---

[21] Mattox, *Augustine*, 60.
[22] Ibid., 60-61.
[23] Ibid., 61.

conflict as well as any promises that are made.[24] Two examples of this concept come to mind. Regarding permissible deception, it is acceptable for an army to conceal its movements and mislead the enemy regarding its intentions. These actions would aid the army and thus, by shortening the conflict, bring a faster end to the war. However, if one commander were to approach the other under a white flag of truce in order to discuss issues such as care for the wounded, then that symbol of truce must be honored.

Jus Post Bellum

Jus post bellum is a Latin term that means, literally, "justice after war." The concept was envisioned after the aftermath of World War I when it became apparent that the Treaty of Versailles, which was supposed to produce a lasting peace, created the conditions that would cause the next war.[25] Even at the time, it was apparent that the treaty was overly punitive towards Germany, thus violating some of the principles discussed above. In order to avoid repeating this mistake, theorists have asserted that procedures need to be put in place that will carry out Augustine's principle of "the establishment of a just and lasting peace."[26] However, the means by which this peace can be achieved are quite complex and, frankly, costly for the victorious nation.

Four key criteria stand out. Incidentally, these criteria feature prominently within this study's *jus post bellum* research. First, the victorious nation needs to shift its focus from occupying enemy territory to rebuilding cities, economic features, and infrastructure.[27] The idea is that a comfortable, prosperous populace is unlikely to seek out another conflict or choose leaders who could be a risk to a lasting peace. A good example of this idea in practice was the Marshall Plan, which rebuilt West Germany after World War II. As a result, West Germany recovered rather quickly and remained independent of influence from the Soviet Union.

There is also a need to render legal justice upon the conclusion of hostilities. War crime trials, such as those that were conducted at Nuremberg after World War II, serve to hold the leaders of belligerent nations accountable.[28] One apparent result of these trials is that they ensure that

---

[24] Ibid., 64-65.
[25] Louis V. Iasiello, "Jus Post Bellum," *Naval War College Review* 57, no. 3 (2004): 38.
[26] Ibid., 39.
[27] Albert W. Klein, "Attaining the Post Conflict Peace Using the *Jus Post Bellum* Concept," *Religions* 11, no. 4 (2020): 174.
[28] Alex J. Bellamy, "The Responsibilities of Victory: *Jus Post Bellum* and the Just

people who commit war crimes are punished and, thus, are an example to others. However, this idea of public justice also serves to prevent victors from executing post-war prisoners without a trial. In other words, those charged with war crimes receive an adequate defense.

A third common *jus post bellum* ideal is the need to provide humanitarian aid. This is a different concept from economic reconstruction in that it addresses the plight of displaced persons or of any citizens who require immediate food or medical attention as a result of the hardships that they experienced during the war. During any war, non-combatants are liable to suffer unjustly through lack of access to food, shelter, and medical care. In order to keep with the principle of reducing suffering, these needs should be met. Of additional note, any war will necessarily displace a portion of the population. The needs of these people must also be addressed.

Finally, there may be a need for the victor to completely reform the government of the defeated state simply because that government can no longer serve the needs of its people. As Gary Bass argues when discussing the need to replace genocidal regimes, governmental structures that cause suffering have lost their right to continue to function.[29] It is important to note, though, that the mindset of the victor is important. There is a vast difference between removing an illegitimate government and installing a puppet regime for the purpose of exerting influence over a nation.

Thesis

The thesis of this work is that Augustine would approve of some, but not all, of the present-day components of *jus post bellum*. One issue is that Augustine wrote under a certain set of circumstances. In his world, he was addressing the new requirements and challenges brought about by what was the world's first nominally Christian government. As a bishop and as a pastor, he believed that he had a duty to respond to this rapidly changing ethical landscape. Another issue is that, unlike the other components of just war theory, *jus post bellum* has a relatively short history.

One major principle of *jus post bellum*, the need to prevent conflict in the future, appears, on the surface, to be as relevant in the present day as it was in Late Antiquity. However, the issue becomes much more complicated when we attempt to precisely define the actions that lead to a permanent

---

War," *Review of International Studies* 34 (2008): 612.
[29] Bass, "Jus Post Bellum," 398.

peace. This is more than merely a theological issue, even though a key difference between the two contexts is that Augustine's definition of "peace" was undoubtedly different from the way "peace" is understood today. Augustine could not have envisioned current post-war activities such as war crime trials and economic reconstruction. The Roman government simply did not perform these actions. Furthermore, even these Postmodern attempts at establishing a permanent peace often sound ethical but tend to produce the opposite of the intended effect. Therefore, discerning exactly which *jus post bellum* activities are truly in accordance with Augustinian principles will be a complex exercise.

Methodology

This study will examine Augustine's thoughts on war, to include the context in which these writings were developed, before tracing these thoughts to the present day. It will then explain the rise of the third pillar of just war theory in the twentieth century. With this overarching context in mind, this work will review *jus post bellum* in practice.

This work will examine four key components of *jus post bellum* across three different post-conflict eras. The three post-conflict eras are the post-World War II reconstruction of Japan and Germany, the American experience in Iraq, and the American experience in Afghanistan. The scope of the study has been limited to these three areas for a variety of reasons. First, Augustine has had the greatest influence on the Western world. Other world cultures have tended to draw their principles of ethics during wartime from other sources.[30] Second, only a large, prosperous nation could have performed many of the *jus post bellum* activities that have taken place since World War II. Economic reconstruction of an entire nation, for example, necessarily requires vast sums of money. Third, each one of these *jus post bellum* eras had different outcomes. Japan, for instance, was successfully rebuilt but Afghanistan appears to have fallen into an even deeper state of disarray.

Even though *jus post bellum* attempts in these three eras had different results, the post-war processes that were utilized during these eras were quite similar. For example, economic reconstruction was a central component of all three eras. This study will examine the success or failure of war crime trials, humanitarian aid, governmental changes, and economic reconstruction. These efforts are well-critiqued by scholars who tend to hold a wide variety of opinions. For example, some find that war crime trials provide necessary

---

[30] Mattox, *Augustine*, 2

justice after a conflict while others assert that these trials only provide an incentive for belligerents to fight well after the outcome has been decided. This diversity of opinion will enhance this study as it will prove that *jus post bellum* research is still in its formative stages.

Within each era of study, the results of these four components will be consolidated and compared to Augustinian thought. Then, this work will conclude with a summation of findings as well as the practical implications of this research. Of additional note, this study will provide numerous indications of areas for further study. Given the very recent conclusion of *jus post bellum* activities in Afghanistan, the data available to this particular field of ethical study has greatly expanded and, thus, lends itself to broad scholarly analysis.

2. AUGUSTINE'S VIEWS ON WAR

The Context of Late Antiquity

Augustine's views on the subject of war were framed by a context that may seem foreign to the Postmodern mindset. The Roman Empire had arisen from pagan origins. Shortly after the transition from republic to empire, the Romans insisted that emperors be regarded as divine. Naturally, this was a problem for the early Christians. Even as Christianity spread, individual Christians seemed to be at odds with a government that was, at best, indifferent to them and, at worst, openly hostile to them. One implication of this dynamic is that, until the time of Constantine, Church leaders did not see a valid requirement for Christians to wage war on behalf of the state. Christians lived within the state but, in some ways, were separate from the state even though many did choose to lead lives of public service. In the opinion of some leaders, however, their duty was to spread the gospel and prepare for Christ's return. In Tertullian's mind, for example, Christians had a clear choice between serving God and serving the emperor and, therefore, could only choose one and, thus, reject the other.[31] For Tertullian, it was not a matter of supporting one master, God, over the state but, rather, following God and rejecting any pagan master.

With Constantine's conversion, though, that dynamic changed suddenly and radically. Indeed, many of Augustine's writings on war address this redefined relationship. Mattox describes this dynamic when he writes, "It

---

[31] Philip Wynn, *Augustine on War and Military Service* (Minneapolis, MN: Augsburg Fortress Publishers, 2013), 35.

seems clear enough that Christians in Augustine's sphere of influence were concerned about the propriety of military service and the justifiability of war, in general, to prompt his repeated, if fragmentary, treatment of the issue."[32] If Augustine's particular fifth-century context could be subdivided into four areas, they would be the rise of the first Christian emperor, the rise of Christianity as the dominant religion, the collapse of Roman authority in the West, and the role of the Roman Christian as a citizen.

A Christian Emperor

In 312, during a civil war between himself and three rival co-emperors, Constantine faced what was supposedly a superior force at the Battle of Milvian Bridge. At some point before the battle, Constantine reportedly had a dream or a vision of Christian religious significance. Opinions vary as to exactly what Constantine saw and when he saw it. According to Eusebius, "About noon, when the day was already beginning to decline, he saw with his own eyes the trophy of a cross of light in the heavens, above the sun, and bearing the inscription, 'Conquer by this.' At this sight, he himself was struck with amazement, and his whole army also."[33] However, it is possible that Eusebius' account is a later embellishment. Lactantius, in his own account, appears to claim that Constantine was told by God to place the sign of Christ over his soldiers' shields in a dream.[34] Like Eusebius, though, he, too, wrote this account well after the battle had taken place. Timothy Barnes takes a more balanced approach, perhaps, when he writes, "Knowing when Constantine became a Christian is difficult to uncover, but it is important that he made the public decision before the Milvian Bridge."[35] These accounts remain relevant, though, because they demonstrate how some Roman Christians perceived their first Christian emperor.

It is important to note that, with Constantine's embrace of Christianity, the Roman Empire was still technically a pagan empire. In fact, the Roman Empire did not fully Christianize until the time of Theodosius at the end of the fourth century. What did change immediately was the empire's

---

[32] Mattox, *Augustine*, 143.
[33] Eusebius of Caesarea, *The Life of the Blessed Emperor Constantine* in *Eusebius: Church History, Life of Constantine the Great, and Oration in Praise of Constantine*, ed. Philip Schaff and Henry Wace, trans. Ernest Richardson (New York, NY: The Christian Literature Company, 1890), 490.
[34] Timothy Barnes, *Constantine: Dynasty, Religion and Power in the Later Roman Empire* (West Sussex, UK: Wiley Blackwell, 2014), 79.
[35] Ibid., 80.

treatment of its Christian subjects. With the Edict of Milan, the persecution of Christians ended, and Church property was restored. At the time, this was more of a political decision made between Constantine and his then co-emperor, Licinius.[36] An obvious impact of this decision is that Christians could freely worship and spread their faith without fear of prosecution. The nature of this worship had changed, however. As Mattox writes, "Christians now needed to pray for the emperor and for the state, now that the emperor was a Christian."[37] For the first time, there was a need for believers to pray for their Christian secular leaders. Undoubtedly, Christians had prayed for their leaders before this point. In Scripture, for example, Paul writes, "I urge that requests, prayers, intercessions, and thanks be offered on behalf of all people, even for kings and all who are in authority, that we may lead a peaceful and quiet life in all godliness and dignity" (1 Timothy 2:1-2). However, now that believers were under the authority of the world's sole Christian ruler, they needed to pray for the continuity of Christian successors as well as for the continued prosperity of this Christian state.

One additional implication of this new level of tolerance is that, with a Christian emperor, it became much more acceptable for Christians to enter into government service. Indeed, there is evidence that Constantine "began to give preference to Christians in appointments to government and military positions."[38] As a result, as Christians increasingly entered the public sphere, there became an enhanced need to positively associate the Christian lifestyle with public service.

Furthermore, by virtue of being a Christian emperor, Constantine was able to intervene in Church affairs for the purposes of preserving public order, as he did during the Council of Nicaea. This is an ability that previous emperors, such as Diocletian, simply did not possess. In fact, as Henderson and Kirkpatrick note, Constantine had previously called together smaller councils because violence between factions was getting out of hand.[39] Based on these actions, it is reasonable to assert that, in the context of Constantinian rule, it was an accepted practice for the sovereign to insert himself in Church affairs when Church matters had an impact on his ability to rule.

---

[36] Barnes, *Constantine*, 95.
[37] Mattox, *Saint Augustine*, 143.
[38] David E. Henderson and Frank Kirkpatrick, *Constantine and the Council of Nicaea: Defining Orthodoxy and Heresy in Christianity, 325 CE* (Chapel Hill, NC: University of North Carolina Press, 2016), 18.
[39] Ibid., 37.

Though Constantine may have been revered by some Christians as an ideal ruler, there are indications that his spiritual life may have been quite complex. For example, he waited until the end of his life to be baptized. Eusebius places a positive emphasis on the choice when he writes, "At the conclusion of the ceremony, he arrayed himself in shining imperial vestments, brilliant as the light, and reclined on a couch of purest white, refusing to clothe himself with the purple anymore."[40] It is possible that Constantine waited out of a sense that he wanted to be forgiven of his sins shortly before his death. This makes sense given that he did not conclude his civil war with his last co-emperor until well after his conversion experience. But this may also mean that, as a ruler, his spiritual life, at times, fell short of the Christian ideal.

This is an important point because Constantine and his successors were not perfect rulers. Constantine's own son, who became Constantius II, was Semi-Arian in faith. Constantine's nephew, Julian the Apostate, attempted to restore paganism within the Empire. Valentinian I, though nominally Christian, spent much of his time engaging in war with Germanic tribes and was known for being a violent, ill-tempered man. By the time of Augustine, therefore, it is apparent that there was a dichotomy between the office of the Christian emperor and the actual character of the person holding the office.

The question was, what should a Christian do when God's chosen sovereign was anything but a Godly man? This issue was, perhaps, most stark during the reign of Julian. Technically, as a direct successor to Constantine, he was a man chosen by God to rule a rapidly Christianizing empire. Yet, Julian openly turned his back on Christianity and attempted, without success, to return pagan worship to its former place. Augustine took a nuanced approach that will have resounding effects in the present-day discussion of *jus post bellum*. According to Mattox, Augustine argued that Christian soldiers were required to obey their sovereign unless that sovereign gave them a command that contradicted God's laws.[41] The idea was that, regardless of the character of the sovereign, his legitimacy was to be respected up to the point where a sovereign orders a soldier to commit an unlawful act. This concept echoes the present-day duty of soldiers, as representatives of the state, to carry out lawful orders but refuse to commit those actions that are now considered to be war crimes.

---

[40] Eusebius, *Constantine*, 556.
[41] Mattox, *Augustine*, 57-58.

A Dominant Religion

In the period before Augustine's life, Christianity evolved from being a persecuted faith to being a faith that was supported by the Roman government. Christianity became the only official religion soon afterward. This kind of upheaval changed the character of the faith. First, with Constantine's decision to support Christians in governmental and military roles, it became socially acceptable to become a Christian. Naturally, it would be reasonable to assume that many of these newer Christians did not have the same level of fervency as did their predecessors. Second, in order to accommodate this influx of new believers, the Church hierarchy needed to grow and firmly establish itself.

These concepts may sound familiar to Postmodern ears, but an aspect of these developments that would be foreign in a present-day Western context is the idea of the sovereign as the *de facto* head of a large, complex state-sanctioned religion. This idea would not have been unfamiliar to the Romans, however. The emperors had held the office of *pontifex maximus* since the days of Julius Caesar. Throughout the long history of the Roman Empire since that time, the emperor had always been revered as a religious leader, whether the faith be traditional Roman paganism, *Sol Invictus*, or Christianity. The idea of the deification of the emperor, whether living or deceased, was an extension of his role as a high priest.

It was natural for Constantine to retain his role as the high priest of the Roman Empire even though he was now a Christian. However, he took a secular, rather than ecclesiastical, view of this role by seeing himself as a ruler who was responsible for protecting the Church and its structures. Everett Ferguson discusses this mindset when he writes, "Constantine felt a responsibility for the religious welfare of his subjects. He spoke of himself as…bishop of the external affairs of the church."[42] In other words, Constantine did not have the same type of religious authority that belonged to, for example, the Bishop of Rome. However, he had a responsibility to protect the religious health of his subjects, both internal and external, using the tools of statecraft. Medieval Europe would more fully define this idea later in its history, but it may be useful to imagine church and state working together to maintain civic and religious harmony.

Until this time, the Romans had always been pluralistic in matters of faith. With few exceptions, such as with Christianity, they accepted the

---

[42] Everett Ferguson, *Church History Volume One: From Christ to Pre-Reformation* (Grand Rapids, MI: Zondervan, 2005), 186.

worship of additional foreign gods so long as these faiths did not interfere with the well-being of an orderly Roman society. The issue with Christianity, from the pagan Roman perspective, was its claim that all other religions were false. This had an effect on the way that Christians fit into the fabric of Roman society. As Robert Wilken writes regarding Emperor Julian's feelings, "Julian...impugned the Christians for deserting the gods. In the Roman world, this charge was not simply a matter of 'our gods' against 'your God.' The gods were part of an entire social world in which Christianity could not be fitted."[43] Therefore, by definition, the presence of a large, growing segment of society that does not conform to societal norms would necessarily be seen as a destabilizing force.

This dynamic was turned on its head between the time of Emperor Constantine and Emperor Theodosius I. While it would have been natural to assume that a Christian emperor would, by definition, attract a large number of new followers to his faith, perhaps the most significant change came when Christianity became the state religion to the exclusion of all other religions during the reign of Theodosius. Now, a large, powerful, organized faith was the only faith allowed in the Mediterranean basin and it had a sovereign with judicial and military powers at its head.

The problem with this dynamic was that Christianity had been forced to grow too fast. Atheist scholar Bart Ehrman describes this growth as a type of rapid increase in baptisms during the fourth century as Christianity became more acceptable.[44] Certainly, the faith of the emperor made Christianity more attractive. However, the message itself, unbound by state restrictions, did lead to genuine mass conversions.[45] As a result of this dynamic, Church structures and the Church hierarchy needed to rapidly expand in order to meet the increased need. As Ehrman states, "The emperor himself claimed allegiance to the church; more people were joining the Christian ranks daily; church buildings were being constructed; members of the elite were starting to convert."[46] As baptisms grew, church attendance grew and, thus, the Church needed to greatly expand its own organizational structures.

---

[43] Robert Louis Wilken, *The Christians as the Romans Saw Them*, 2nd ed. (New Haven, CT: Yale University Press, 2003), 201.
[44] Bart D. Ehrman, *The Triumph of Christianity: How a Forbidden Religion Swept the World* (New York, NY: Simon & Shuster, 2018), 211.
[45] Ibid, 211-212.
[46] Ibid., 212.

To make matters more challenging, when Christianity became the state religion, it had to present itself as a stabilizing force, which meant that it needed to appear to be unified. As Ferguson notes, "Church-state relations underwent a paradigm shift, now requiring the definition of the competence of a Christian empire. The church found itself largely unprepared for the change from a persecuted church to a favored church."[47] The result is that, at a time when Christianity was expected to present itself as a unified faith for the sake of imperial stability, its very acceptance as a mainstream religion produced the opposite effect.

By the time of Augustine, the Church and the Roman government were involved in the persecution of pagans and heretics, though they approached the issue from different angles. The state sought to preserve civic order and harmony; the Church sought to establish biblical orthodoxy. The end result was that the state stepped in, when deemed necessary, to quell the various heretical movements that seemed to threaten the integrity of the Church.

In an Augustinian context, the Donatist controversy provides, perhaps, the most noteworthy example of church-state involvement in religious matters. Briefly, during the long periods of Roman persecution of Christians, many Christians suffered martyrdom while others sought to preserve their lives and property by taking actions such as taking part in the worship of the emperor, renouncing their faith, and handing over their copies of sacred works. When Christianity was suddenly transformed from a persecuted religion to an approved and state-supported religion, the Church, from an ecclesiastical perspective, needed to decide what to do with those Christians who had, in one way or another, renounced their faith.

In a sense, the Donatist controversy revolved around issues of Christian mercy and Christian purity. Should believers be forgiven and allowed back into the fold now that it was safe to be a Christian? What should be done with Christian leaders who had failed the test of martyrdom? The Donatists took the position that those who had failed in their faith could not rejoin the Church. Furthermore, any sacraments that were administered by failed clergy, in their view, were invalid. The orthodox position, which was held by Augustine, argued that these types of sins could be forgiven. As for the administration of sacraments, they argued, all men were sinners, including the clergy. Therefore, sacraments were provided by the Church and not by a fallible man and thus remained valid.

---

[47] Ferguson, *Church History*, 186.

This issue arose shortly before Constantine declared Christianity to be legal and quickly became a problem for Church unity. Donatists established their own churches and their own clergy. In a sense, they represented a parallel faith to the Catholic Church. In response to the growing schism, Constantine used the powers of the state to persecute the Donatists. However, since the Donatists were, by definition, ardent Christians who embraced martyrdom, this persecution did nothing to quell their rise in popularity. As Raymond Van Dam writes, "At first, he imposed legal penalties, including the confiscation of churches. In the long run, this sort of intolerance boomeranged because the Donatists could argue that the opposition of an emperor, even a Christian emperor, proved they were the authentic heirs of the martyrs."[48] Faced with few remaining options, Constantine ceased his persecution and, as a result, the problem festered until the time of Augustine's ministry.

By the early fifth century, the majority of Christians in North Africa were Donatists.[49] In part, their success was due to their cultural appeal. Ferguson writes, "The Donatists' moral rigor, ethnic identification with the native populations of North Africa, and their appeal to the fathers of the North African church (Tertullian and Cyprian), all gave Augustine a hard job."[50] From his pulpit and from his desk in Carthage, Augustine fought a theological battle against what was, in his locality, a majority faith. It is an unfortunate fact, though, that, shortly after Augustine's lifetime, the political situation in the region changed dramatically. In the final years of his ministry, as Roman authority collapsed, the Vandals, who were Arian Christians, invaded North Africa. They held it until a brief reconquest by the Eastern Roman Empire many years later followed by the Arabs shortly afterward.

The Collapse of Roman Authority

When speaking of the collapse of the Roman Empire, it is best to carefully nuance what is meant by the term "collapse." The Roman Empire had actually split into thirds shortly before the time of Constantine during the Third Century Crisis. His predecessor, Diocletian, finally reformed the Roman Empire, but in doing so, dramatically changed its character by splitting rule of the state into four regions. During a series of civil wars, Constantine united the Roman Empire under one rule but, eventually, the state would be permanently divided into the Eastern Roman Empire and the

---

[48] Raymond Van Dam, *The Roman Revolution of Constantine* (New York, NY: Cambridge University Press, 2008), 264
[49] Ferguson, *Church History*, 274.
[50] Ibid.

Western Roman Empire. In the East, rulership under an emperor in Constantinople would continue until AD 1453. In the West, a series of military defeats, economic crises, and barbarian incursions caused the state to devolve into a group of independent kingdoms by AD 476. During this process, two notable events had a significant effect on Augustine's ministry.

The most important event was the sack of Rome in AD 410 by Alaric's Visigoths. By this point, the capital of the Western Roman Empire had already been moved to the more defensible city of Ravenna, and Rome, in a sense, no longer held the same level of political significance that it had maintained in previous years. However, this sack was the first time that the city had been successfully occupied since it was taken by the Gauls in 390 BC. Therefore, it was a serious blow to morale in the Western Roman Empire and to confidence in imperial authority.

Interestingly, though, the sacking of Rome took on an unexpected character. Peter Heather writes, "By all accounts, there followed one of the most civilized sacks of a city ever witnessed. Alaric's Goths were Christian and treated many of Rome's holiest places with great respect."[51] This is not to say that the sacking was completely bloodless or that the Visigoths chose not to plunder the city. However, in comparison to previous Roman actions, such as the conquest of Carthage during the Third Punic War, or other actions, such as the barbarian invasion of Roman Africa, it was a relatively peaceful affair.

The fact that the Visigoths created places of sanctuary that were free from violence is especially noteworthy. Heather states, "The two main basilicas of St. Peter and St. Paul were nominated places of sanctuary. Those who fled there were left in peace, and refugees from Africa later reported with astonishment how the Goths even conducted certain holy ladies there."[52] During his ministry and within his writings on the subject of lawful warfare, Augustine will take note of this phenomenon, especially in comparison to Roman atrocities in past centuries.

A second major political event during Augustine's lifetime was the Vandal invasion of North Africa. Barbarian tribes had been crossing the frontier of the Western Roman Empire since roughly AD 376. Roman responses were varied. In some cases, such as with the Franks, tribes were allowed to set up semi-autonomous kingdoms in imperial territory. In other

---

[51] Peter Heather, *The Fall of the Roman Empire: A New History of Rome and the Barbarians* (New York, NY: Oxford University Press, 2006), 227.
[52] Ibid.

cases, such as with the Vandals, tribesmen carved out their own territory in areas where imperial political and military strength was at its weakest.

At the end of Augustine's life, the Vandals were laying siege to Hippo. Unlike the situation in Rome, though, the event took on a much more violent nature. Heather writes, "While Geiseric's main army got on with the business of besieging, some of his outlying troops, lacking credible opposition, spread out across the landscape. Leaving destruction in their wake, looting the houses of the rich and torturing the odd Catholic bishop, they moved further west towards Carthage."[53] Whereas at one time, a pair of Roman legions stationed in North Africa would have been able to prevent the incursion, neither the army nor the government had the ability to protect Roman citizens.

Unsurprisingly, the Roman army's deterioration ran concurrently with the worsening political situation. During the fourth century, the army was unable to handle the influx of barbarian tribesmen that were being pushed westward due to the movement of the Huns. The Western Roman Empire, in particular, was required to protect vast borders along the Rhine and the Danube. The Roman army was stretched so thin that, by 401, troops had to be withdrawn from the Rhine and Britain, and a large number of barbarian tribesmen had to be inducted into the regular Army.[54] Even changing its strategy from holding tribesmen at its borders to engaging in a more elastic defense made little difference. Much of Gaul and Spain became permanently occupied by foreign invaders and, eventually, Britain was completely abandoned.

When Alaric sacked Rome in 410, he did not do so after defeating a vast Roman army at the peak of its capabilities. Tribesmen were such a critical part of the army by this point that even its commanders were often of foreign origin. One such commander, Stilicho, was crucial to preserving the integrity of the Western Roman Empire's borders but, upon his death, his army had little loyalty to the Roman state. As Rich and Shipley note, "The army became completely ineffective when after the death of Stilicho its federates, said to have been 30,000 strong, deserted to Alaric.[55] Therefore, when Rome was sacked for the first time since the early days of the Republic, there was no army capable of preventing an enemy force from entering the city.

---

[53] Heather, *Roman Empire*, 271.
[54] John Rich and Graham Shipley, eds., *War and Society in the Roman World* (London, UK: Routledge, 1993), 266.
[55] Graham and Shipley, *War and* Society, 267.

The situation was different in North Africa, where the Vandals invaded with the assistance of inept or corrupt Roman officials. The tribesman had actually moved through Gaul and Spain before staging a naval invasion of North Africa. There was a Roman army that was, theoretically, large enough to oppose them. Heather writes, "Boniface, the count of Africa, had at his disposal 31 regiments of field army troops (a minimum of 15,000 men) as well as another 22 units of garrison troops (at least 10,000 men) distributed from Tripolitania to Mauretania."[56] Given the inherent weaknesses of a slow, unprotected naval invasion, the Romans should have been able to prevent the Vandals from crossing the Strait of Gibraltar.

However, a lack of communication caused by the erosion of imperial authority allowed this invasion to take place unopposed. The issue was that, while the Roman army in North Africa had a sufficient number of soldiers, the Vandals conducted their initial landings in an area that was comparatively weakly defended by a small garrison that was not under Boniface's command. As Heather notes, "The main job of garrison troops had always been to police the comings and goings of nomads, and it must be extremely doubtful whether they were really up to set-piece confrontations with Geiseric's battle-hardened force."[57] Of course, under normal circumstances, the local commander could have established a coordinated defense, but such coordination with other elements of the Roman army never occurred. As a result, the Vandals gathered their strength in North Africa and were eventually able to occupy Carthage as well as Augustine's city of Hippo.

The Role of the Christian Citizen

As both the Church and the state changed dramatically following the events of the Third Century Crisis, the role of the Christian necessarily needed to change. Before the time of Constantine, it was normal for a Christian to be quietly involved in both the government and in the military. Once the Roman Empire became the world's sole Christian state, a believer's civic duties began to include taking actions that would preserve the state.

Before the time of Constantine, it would be fair to say that Christians were able to serve but that this service did not receive overt approval from either church leaders or from the government. The reasons were quite complex. Tertullian, for example, opposed any type of Christian military service. Wynn writes, "Thus absolute rejection is echoed years later in his *De Corona*, where Tertullian again questioned 'whether military service is proper

---

[56] Heather, *Roman Empire*, 268.
[57] Ibid., 270.

for all Christians' and clearly concluded that it was not, even coming close to advocating desertion for anyone who converted while in the service."[58] However, Tertullian did not feel that the defense of a pagan state was inappropriate for a Christian. The issue was more theological in nature. Wynn writes, "Tertullian based his opposition to Christian military service not on any ethical objection to killing in war per se but on the consideration that it was impossible for soldiers to avoid any contact with pagan religious rites associated with Roman army life."[59] This was a valid point. For the entire history of the Roman army, going back to the days of the Early Republic, the soldiers participated in a variety of pagan rituals, such as observing the flight of birds or sacrificing an animal to determine whether or not the day was favorable for battle. It is natural that the clergy would desire to keep believers away from these types of practices.

Origen took a different approach by seeing Christians as being set apart within the Roman Empire as a separate type of priestly class. Wynn notes, "Origen in his *Contra Celsum* replied that, as with the pagan priesthood, Christians had to keep their hands undefiled from bloodshed so that their prayers for the emperor and his armies might be more effective."[60] Again, the idea was not so much that it was wrong to defend the state, but that Christians had a higher duty to maintain their purity.

However, there is evidence that, despite these exhortations, Christians chose to quietly serve in the Roman army. As Wynn states, "The recent archaeological find of a Christian place of worship within the fort of Dura Europos in Roman Mesopotamia indicates that by the early third century not only were there significant numbers of Christians in the army, but that to some extent their presence had received at least tacit 'official' recognition."[61] In other words, at the lower levels of command, the presence of Christians within the army was at least acknowledged and tolerated. Furthermore, the fact that a large number of Christians were able to serve means that, at least in some parts of the Empire, fear of persecution by the government or of chastisement by their spiritual leaders was minimal.

Persecution within the army was not absent, however. Once Christians became visible within the imperial court, their presence was an issue. As Wynn states, "A sign of pagan nervousness over the enhanced visibility of Christians in the military's highest ranks is the fact that Diocletian

---

[58] Wynn, *Augustine on War and Military Service*, 36.
[59] Ibid.
[60] Ibid., 36-37.
[61] Ibid., 38

was persuaded to purge Christians from the army."[62] Therefore, it can be assumed that the presence of Christians was less tolerated in Rome and more accepted in the provinces.

This dynamic changed when Constantine embraced Christianity. It is important to realize that he once was one of four co-emperors before he established unified control over the Roman Empire. Between the time that he announced his conversion and the time that he achieved full rulership, he was the only co-emperor that Christians could support. As Wynn notes, "Christians throughout the empire were deeply involved and deeply invested in the outcomes of the civil wars of 313 and 324. In the one instance, they had endured as much as a decade of the worst persecution the church had ever suffered; in the other, a persecution was begun or feared."[63] In the recent past, Christians faced serious persecution under Diocletian. Later, while it was true that one co-emperor joined Constantine in issuing the Edict of Milan, Constantine's other co-emperors were decidedly less pro-Christian. Therefore, it was natural for the Church to closely align itself with him.

Constantine's embrace of Christian participation in his army had two lasting effects that would certainly influence Augustine's views on war. First, even though Constantine simply tolerated Christian presence in the Roman Empire, he appeared to only favor Christian worship in the army. All pagan symbols and practices were removed from the army and replaced with Christian counterparts.[64] Furthermore, the army gained a distinctly Christian character. As Wynn notes, "Certainly the best example of an effective Constantinian symbol in this regard is the *chi-rho* monogram he had ordered to be put on his soldiers' shields before the battle of the Milvian Bridge. In addition, Sunday was set aside as a day of worship when Christians in the ranks had to leave to attend church."[65] It is possible that he performed this action in order to preserve the allegiance of soldiers within his own ranks. Indeed, only a few years prior, Christians in Diocletian's army were suddenly forced to hide their faith out of fear of retribution, thereby creating a disciplinary problem.[66] Given the rapid Christianization of military practices, it would also be fair to argue that the presence of pagan soldiers in Constantine's army began to decline. Now, given the new characteristics of the army, a case could be made within the Church that the Christian soldiers were holy warriors under the command of God's ruler. As Wynn writes,

---

[62] Ibid.
[63] Ibid., 70.
[64] Ibid., 59-60.
[65] Ibid., 60.
[66] Ibid., 58.

"Christian supporters also began to frame the events of the reign, and the emperor himself, within the contours of ongoing salvation history."[67] The emperor and the Church were united in an effort to spread the faith to mankind.

Second, Constantine made a distinction between the priestly class of Christians and those who could join the army, thus undercutting one of the previous arguments for Christian pacifism. Wynn states, "Constantine had, in effect, acceded to Origen's argument of over a half-century earlier, but with the crucial difference that whereas Origen had apparently argued that *all* Christians should have the same immunity from military service as a pagan priesthood, Constantine intended that privilege to apply *only* to Christian clergy."[68] The end result was that the clergy could support the sole Christian emperor while the laity were free to fight for him. Furthermore, this dynamic of church-state relations would last well past the Western Roman Empire's transition into a series of feudal kingdoms.

Augustine's Writings

The City of God

When Augustine wrote *The City of God*, it was in response to a genuine crisis in Christianity. Since the Christianization of both the Roman army and the Roman Empire, the Western portion of the Roman Empire had begun an irreversible decline. Repeated barbarian incursions were chipping away at Roman territorial integrity. Roman armies were often led by foreign military strongmen who had less than full allegiance to the state. Then, most notably, Rome was sacked in AD 410 for the first time in roughly seven hundred years. It was understandable that pagans would blame Christians for these events as, in their mind, a failure to worship pagan gods as well as an insertion of Christian ideals into society were contributing factors. The Christian God must be inferior to the pagan gods, opponents believed, because Rome was safe when citizens worshipped a wide pantheon of deities. Furthermore, these critics felt that Christian virtues of love and forgiveness were inferior to traditional Roman values because they only served to weaken the moral fabric of the populace.

In Augustine's mind, the sack of Rome only served to confirm his feeling that, while the Rome Empire had turned to Christianity, its behavior

---

[67] Ibid., 71.
[68] Ibid., 59.

beforehand almost assured that it would experience political and military upheaval. As Mattox writes, "However personally Augustine may have bemoaned Rome's fall, it demonstrated conclusively to him that the city of Rome and the kingdom of God could not possibly be one and the same."[69] Of course, a natural reply from the pagans would be that Rome had been a Christian state for over a century, thus giving the faith plenty of time to adjust the morals of the state.[70] Augustine's response has interesting implications for this study. According to Mattox, he states that "the fact that one claims citizenship in an ostensibly Christian state neither makes one truly Christian nor truly just."[71] This idea indicates that Augustine was not naïve enough to believe that a Christian state would always be composed of faithful people led by a righteous ruler. He recognized that even the best human institutions are subject to failure.

Augustine penned *The City of God* as a means of answering the charges that Christianity was responsible for this event. He writes, "My dear Marcellinus: This work which I have begun makes good my promise to you. In it I am undertaking nothing less than the task of defending the glorious City of God against those who prefer their own gods to its Founder."[72] By making a distinction between an earthly city and a heavenly city, Augustine sets up a biblical argument that Christians should indeed take on holy virtues and seek to perform the duties that God has given them. In other words, the earthly city has always been corrupted, but Christians look forward to communion with God in a perfect, heavenly city.

Augustine makes this point by examining Roman history and pointing out a fallacy in the pagan argument. Namely, from the days of the republic until the time of Trajan, Rome had always expanded, but it had experienced more than its share of tragedy and moral corruption along the way. He writes, "The barbarous civil wars which, on the admission of their own historians, were more vindictive than all foreign wars on record, and which not only plagued the republic but utterly ruined it, all occurred many years before the coming of Christ."[73] During the various wars between Marius and Sulla, between the First Triumvirate, and between the Second Triumvirate, pagans were not blaming the bloodshed on a lack of religious

---

[69] Mattox, *Augustine*, 24.
[70] Ibid., 24-25.
[71] Ibid., 25.
[72] Augustine, *The City of God: Books I-VII*, 17.
[73] Ibid., 184.

virtue. The same could be said about Roman military defeats, which were numerous, before the time of Constantine.

Augustine traced Rome's moral decline back to its rise to status as a power in the Mediterranean during the Punic Wars. He states, "Romans now sought for honors and fame, not by the older ways, but by the crooked paths of chicanery and guile."[74] The idea was that expressions of traditional Roman virtue, such as the story of Cincinnatus leaving his fields to rule the Republic only to return to the fields once the time of crisis had passed, had become a type of legendary distant memory centuries before the rise of Christianity.

Furthermore, in Augustine's mind, it is foolish to attribute previous successes to pagan gods and present failures to the one Christian God. God has always existed and has allowed both fortune and misfortune to fall on those who did not worship Him. He writes, "The one true God, who never permits the human race to be without the working of His wisdom and His power, granted to the Roman people an empire, when He willed it and as large as He willed it."[75] The idea is that God has always been working within the earthly city in order to accomplish His purposes. To believe that God would act as a pagan god, distributing earthly favors to those who worship Him, would be to misunderstand His character.

While making these arguments, Augustine said a great deal about the proper conduct of war. Critically, Augustine finds Christian virtue among the "barbarians" who sacked Rome because they demonstrated restraint that the pagan Romans had typically lacked. When the Goths sacked Rome, they actually allowed people to enter places of worship for their own safety. He writes, "Even these ruthless men, who in other places customarily indulged their ferocity against enemies, put a rein to their murderous fury and curbed their mania for taking captives, the moment they reached the holy places."[76] It is important to note that Alaric's Goths were Arian Christians who, while technically being of the same faith as the Roman Christians, were considered heretics. These tribesmen made an effort to keep violence to some kind of a minimum. In comparison with Roman behavior at the end of the Third Punic War, in which the Roman army systematically destroyed Carthage, their level of restraint was noteworthy.

Several principles inherent in present-day just war theory can be found in Augustine's account of the sack of Rome. First, with few exceptions,

---

[74] Ibid., 269.
[75] Ibid., 291.
[76] Ibid., 19.

non-combatants are to be exempt from fighting between soldiers. Those who sought or were provided with sanctuary were, by definition, not involved in any military action against the Goths. It is also noteworthy that the Goths made an effort to provide some noncombatants with assistance. Second, under just war doctrine, combatants should avoid unnecessary destruction of property. It is true that the Goths caused a certain level of destruction and, thus, they do not serve as exemplars for this aspect of *jus in bello*, but, compared to their peers, their actions were somewhat muted.

In discussing conduct during a war, Augustine was also careful to note the reasons why a Christian should be authorized to take violent action. In part, he uses a biblical argument. He writes, "The same divine law which forbids the killing of a human being allows certain exceptions, as when God authorizes killing by a general law or when He gives an explicit commission to an individual for a limited time."[77] In other words, there was no requirement for Christians to be pacifists since God did allow killing under certain circumstances, such as when David or Joshua waged war.

Notably, those who have been commanded by God to wage war in fulfillment of His purposes do so without ulterior motives and they do so in an effort to correct some injustice. Under Augustinian thought, a Christian who serves in an army is also exempt from any restrictions against killing because the soldier is a representative of a God-ordained ruler. He states, "Since the agent of authority is but a sword in the hand, and is not responsible for the killing, it is in no way contrary to the commandment, 'Thou shalt not kill,' to wage war at God's bidding, or for the representatives of the State's authority to put criminals to death, according to law or the rule of rational justice."[78] This presents a very important distinction. For killing to be acceptable, even when done in a manner that is sanctioned by God and is performed for the correct purposes, the soldier must be in a state-sanctioned military organization and under the authority of the ruler.

Within *The City of God*, Augustine also writes about some of the conditions that should be present before a ruler decides to go to war. These ideas will later be considered part of the concept of *jus ad bellum*. First, in his mind, any ruler who contemplates commencing hostilities will do so only with a great degree of hesitation. Augustine writes, "I know the objection that a good ruler will wage wars only if they are just. But surely, if he will only remember that he is a man, he will begin by bewailing the necessity he is

---

[77] Ibid., 53.
[78] Ibid.

under of waging even just wars."[79] By referring to the ideal of a just ruler, Augustine presupposes that the ruler who contemplates initiating a conflict holds that position of rulership legitimately and morally. The ruler is the head of a state and, as such, has been allowed by God's grace to hold that position.

Naturally, not every ruler is just, and in fact, some wage war without a sense of the tragedy that will befall both nations. Augustine makes this clear elsewhere in *The City of God*. For those who contemplate warfare without a sense of conscience, he writes, "Any man who will consider sorrowfully evils so great, such horrors and such savagery, will admit his human misery. If there is any man who can endure such calamities, or even contemplate them without feeling grief, his condition is all the more wretched for that."[80] This lack of conscience does not necessarily detract from the legitimacy of the ruler. In fact, Augustine had made it clear that God allowed many immoral men to lead to Roman Empire. But the fact that wars must be conducted as a means of last resort, and then only in deep contemplation of the tragedy that will take place, means that, under Augustinian thought, wars should not commence unless the goal is to cure some injustice. In fact, when the ruler, after considering his options, chooses to commence a war, it should be with the eventual goal of preventing suffering.

To put this another way, war will take its toll in lost lives and human suffering, but a higher degree of suffering would occur if the ruler chose to refrain from warfare. Augustine writes, "A good man would be under compulsion to wage no wars at all, if there were not such things as just wars. A just war, moreover, is justified only by the injustice of an aggressor; and that injustice ought to be a source of grief to any good man, because it is human injustice. It would be deplorable in itself, apart from being a source of conflict."[81] This statement also implies that the just ruler should not use conflict as a means to achieve material gain either for himself or for the state. The sole purpose for beginning war should be to end injustice, not to achieve a gain in raw goods, material, land, or military glory.

Critically for the later idea of *jus post bellum*, Augustine finds that peace should be the end goal of military conflict. He states, "The purpose even of war is peace. For, where victory is not followed by resistance there is a peace that was impossible so long as rivals were competing, hungrily and

---

[79] Augustine of Hippo, *The City of God: Books XVII-XXII*, trans. Gerald G. Walsh and Daniel J. Honan (Washington, D.C.: The Catholic University of America Press, 2008), 206-207.
[80] Ibid., 207.
[81] Ibid., 207.

unhappily, for something material too little to suffice for both. This kind of peace is a product of the work of war, and its price is a so-called glorious victory."[82] Importantly, he seeks a lasting peace. It is not enough that war end, the injustice cease, and that the victor depart. The very conditions that caused the injustices, such as rivalries or competition over scarce goods, must also be corrected in order to prevent future wars.

To conclude on a theological note, Augustine was responding to a controversy over the role of the Christian in the Roman Empire and to a sentiment that Christians were responsible for Rome's recent collapse. In making a differentiation between the fallible, earthly city and God's city, Augustine is also making an argument that the purpose of Christian leadership and the role of the Christian is to spread the gospel so that as many people as possible could look away from the earthly city and towards a future eternal life with God. The peace that Augustine desires should, ideally, allow people time and space to come into a relationship with God, thereby finding eternal peace. The righteous ruler, as a servant-leader of the state who is supported by both the believers and the Church, should keep this high ideal in mind as he seeks to establish tranquility both within and outside his domain.

Reply to Faustus the Manichaean

At first glance, it may not seem that Augustine's *Reply to Faustus* was a treatise on warfare. After all, *The City of God* was intended, in part, to explain why it was acceptable for Christians to join the Roman army. In contrast, *Reply to Faustus* was more about addressing attacks on the historicity of the Old and New Testament accounts. However, in addressing Faustus' issues with the figures presented within the Old Testament, Augustine says a great deal about the subject of state-sponsored war.

One of the many charges leveled against Moses was that he was a man of war who "committed murder, and plundered Egypt, and waged wars, and commanded, or himself perpetrated, many cruelties."[83] It is true that Moses did wage wars as he sought to bring the Israelites to Canaan, such as when he led the Israelites against the Amalekites as described in Exodus. The author states, "Amalek came and attacked Israel in Rephidim. So Moses said

---

[82] Ibid., 419-420.
[83] Augustine of Hippo, *Reply to Faustus the Manichaean* in *A Select Library of the Nicene and Post-Nicene Fathers of the Christian Church*, Volume IV, *St. Augustine: The Writings Against the Manichaeans and Against the Donatists*, trans. Richard Stothert, ed. Philip Schaff (Buffalo, NY: The Christian Literature Company, 1887), 274.

to Joshua, 'Choose some of our men and go out, fight against Amalek. Tomorrow I will stand on top of the hill with the staff of God in my hand" (Exodus 17: 8-9).[84] Note, though, that while Moses is acting as a leader, the battle is taking place with God's full support and encouragement. Moses is acting as, in what would be Augustine's opinion, a righteous sovereign.

It may also be possible to argue that Moses was a proponent of genocide. The biblical account of Moses' exhortation to the Israelites as they prepared to enter the Promised Land could be interpreted as a present-day description of war crimes. The author states, "As for the cities of these people that the Lord your God is going to give you as an inheritance, you must not allow a single living thing to survive. Instead, you must utterly annihilate them – the Hittites, Amorites, Canaanites, Perizzites, Hivites, and Jebusites – just as the Lord your God has commanded you" (Deuteronomy 20:16-17). However, the commandment to clear away any semblance of pagan worship in the Promised Land was more about ensuring that the Israelites could remain holy and, thus, serve as God's representatives to the other nations. Furthermore, the tribes that were to be extinguished had been considered God's enemies for a considerable length of time by this point.

Augustine finds that Moses was acting properly by seeking to wage wars against the occupiers of the Promised Land. He writes, "You profess to accuse Moses of doing wrong, while in fact you envy his success. There was no cruelty in punishing with the sword those who had sinned grievously against God. Indeed, Moses entreated pardon for this sin, even offering to bear himself in their stead the divine anger."[85] In keeping with Augustine's view of the righteous ruler, Moses was not a belligerent man but, rather, engaged in war reluctantly. He made every attempt to avoid bloodshed before submitting to God's will.

In an attempt to explain how Moses can instigate conflict and remain morally blameless, Augustine explains that, as God's agent, he cannot be held responsible for doing as he was commanded. He writes, "It is therefore groundless calumny to charge Moses with making war, for there would have been less harm in making war of his own accord, than in not doing it when God commanded him."[86] In other words, if Moses had chosen to disobey

---

[84] Unless otherwise noted, all biblical passages referenced are in the *New English Translation*.
[85] Augustine, *Reply to Faustus*, 312.
[86] Ibid., 303.

God, then he could have expected serious consequences, both for himself and for the Israelites.

As to the question of whether or not God is acting in a moral manner when He requires Moses to engage in warfare, Augustine implies that God uses conflict as a means to correct some type of injustice. He states, "Thus in all the things which appear shocking and terrible to human feebleness, the real evil is injustice; the rest is only the result of natural properties or of moral demerit."[87] Returning to some of the points made in *The City of God*, it is reasonable to be horrified by the impact of war, but it is important to understand that the very purpose of righteous, God-sponsored warfare is to correct ongoing injustice when other means have been exhausted. In the case of the enemies of Israel, the pagan tribes had either presented a problem for the Israelites in the past or appeared to be a serious threat to the future integrity of the nation.

Interestingly, Augustine also finds that Jesus approved of war, when appropriate. On the occasion of Jesus' arrest, John writes, "Then Simon Peter, who had a sword, pulled it out and struck the high priest's slave, cutting off his right ear" (John 18:10). As the account continues, Jesus tells Peter to cease the attack and put his sword away. However, the very fact that Peter had been carrying a sword means that Jesus had been allowing him to carry it for some time. The implication is that Peter had the capability to perform a violent act, but only when God specifically allowed him to do so. As Augustine states, "Doubtless, it was mysterious that the Lord should require them to carry weapons and forbid the use of them. But it was His part to give the suitable precepts, and it was their part to obey without reserve."[88] It is true that Peter was a single man and not an agent of a state. Due to his close companionship with Jesus, though, the principle of engaging in violence when required by God still applied.

In making this argument, Augustine also dispels the idea that the God of the New Testament was unlike the God of the Old Testament because He absolutely prohibited warfare. In fact, the New Testament seems to provide tacit support for the military profession, so long as soldiers conduct themselves in a restrained manner. For example, Luke writes that when John the Baptist baptized a group of Roman soldiers, he said, "Take money from no one by violence or by false accusation and be content with your pay" (Luke 3: 14). Augustine finds that John must have supported the idea of Romans being soldiers even though, by virtue of the fact that he met

---
[87] Ibid.
[88] Ibid., 302-303.

them, they were part of an occupying army. He writes that John could have told them to "throw away their arms" but instead knew that the soldiers were there to "defend the public safety."[89] Of note, the Roman soldiers were not under direct divine authority. Returning to ideas presented in *The City of God*, though, they were well-behaved peacekeepers who were under the authority of an emperor whom God allowed to remain in authority over them.

Letters

Augustine was a prolific writer and, in addition to classic works such as *The City of God*, he wrote a vast number of letters, often in an effort to provide mentorship in an era when written communication, though laborious, was to best means to communicate over great distances. For example, as Edward Smither writes, "We have noted how Augustine sent several letters accompanied by books intended to resource the clergy."[90] Some of his letters addressed the subject of the just conduct of war. Notably, for the doctrine of *jus post bellum*, Augustine often speaks about the need to establish a lasting peace.

In a letter to a man named Darius, who was a Christian and held a high station with the Roman government, Augustine advised him to seek peace whenever possible. However, he makes an interesting distinction between the roles of the soldiers and the role of the leader. Regarding the duty of the soldier, Augustine writes, "But it is a higher glory still to stay war itself with a word, than to slay men with the sword, and to procure or maintain peace by peace, not by war. For those who fight, if they are good men, doubtless seek for peace; nevertheless, it is through blood."[91] Those who fight, if they are righteous, moral agents of the state, do not desire to conduct war but are held blameless in their conduct if they are ordered to do so. The soldier does not gain glory by war since both the soldier and the ruler prefer that conflict be avoided. However, Augustine places a higher duty upon the righteous leader who makes the initial decision to enter into conflict. He writes, "Your mission, however, is to prevent the shedding of blood. Yours, therefore, is the privilege of averting that calamity which others

---

[89] Ibid., 300-301.
[90] Edward L. Smither, *Augustine as Mentor: A Model for Preparing Spiritual Leaders* (Nashville, TN: B&H Academic, 2009), 212.
[91] Augustine of Hippo, *Letters of St. Augustine* in *A Select Library of the Nicene and Post-Nicene Fathers of the Christian Church*, Volume I, *The Confessions and Letters of St. Augustine with a Sketch of His Life and Work*, trans. J.G. Cunningham, ed. Philip Schaff (Buffalo, NY: The Christian Literature Company, 1886), 581-582.

are under the necessity of producing."[92] Unlike the soldier, the ruler has a duty to perform every reasonable action that could solve the injustices that are present without necessitating bloodshed.

In a letter to Boniface, Augustine encourages the reader to not only refrain from engaging in war for personal gain but to also ensure that the structures of the world remain available for the good of mankind. He writes, "This, however, does not affect your obligation to love God and not to love the world, to hold the faith steadfastly even in the cares of war, if you must still be engaged in them, and to seek peace; to make the good things of this world serviceable in good works, and not to do what is evil in laboring to obtain these earthly good things."[93] It is possible that this idea can carry over into the *jus post bellum* idea of economic reconstruction. This is not to say that Augustine is speaking about economic reconstruction on a twentieth-century scale in this passage. However, it is reasonable to believe that he prefers that peace-producing elements of the economy either remain untouched by war or be restored after the commencement of peace. Furthermore, from a theological standpoint, Augustine makes a distinction between the love for the world that produces a desire for personal gain and a love for the world that seeks a temporal type of peace. In the former, an unrighteous ruler seeks to plunder another nation for his own gain. In the latter, the righteous ruler loves those things of the world that help maintain the peace of God.

In a letter to Marcellinus, Augustine finds that the just ruler who wishes to re-establish peace should remove those things that provoke men or cause them to sin. He writes, "And in mercy, also, if such a thing were possible, even wars might be waged by the good, in order that, by bringing under the yoke the unbridled lusts of men, those vices might be abolished which ought, under a just government, to be either extirpated or suppressed."[94] This principle is somewhat open-ended as a great number of things could lead the citizens of the defeated nation into sin. As a general statement, though, it would be fair to say that the victorious ruler should seek to re-establish the presence of law enforcement entities as well as courts of law within the defeated nation as soon as practicable.

In another letter to Boniface, Augustine again stresses the need to maintain peace and to engage in warfare solely for the reason of reestablishing peace. He writes, "Peace should be the object of your desire; war should be waged only as a necessity and waged only that God may by it deliver men

---

[92] Augustine, *Letters*, 581-582.
[93] Ibid., 576.
[94] Ibid., 486.

from the necessity and preserve them in peace."⁹⁵ In the same letter, he makes an interesting remark about the need for mercy during wartime. Augustine writes, "Let necessity, therefore, and not your will, slay the enemy who fights against you. As violence is used towards him who rebels and resists, so mercy is due to the vanquished or the captive, especially in the case in which future troubling of peace is not to be feared."⁹⁶ This component of the letter alludes to a number of concepts that will appear later within just war theory.

In a present-day context, military leaders require that soldiers show humanity towards captured enemy combatants. Once an enemy no longer has the will or the capability to resist, then there is no further need to engage in violence. The former enemy can become a prisoner of war, but the victorious army must treat him well. This treatment includes any actions that prevent unnecessary suffering such as offering him food, water, and medical attention. The overarching idea is that the ethical army should use violence out of necessity and not out of a desire to cause destruction. The captor can certainly interrogate the captive and later house him in spartan facilities until he is repatriated, but to deny a minimum of comfort to the captive or to torture the captive would be unlawful.

Elsewhere, Augustine reminds the reader that soldiers are to be regarded as citizens who are under many of the same strictures as Christians who are not in the army. After discussing the New Testament account of the baptized soldiers who were encouraged to perform well rather than lay down their arms, he says, "Let those who say that the doctrine of Christ is incompatible with the State's well-being, give us an army composed of soldiers such as the doctrine of Christ requires them to be."⁹⁷ The idea is that, just like anyone in authority, such as a parent, master, or judge, the soldier should conduct himself in a way that sacrifices selfish needs for the greater good.⁹⁸ The fact that this standard applies equally to civilians and soldiers alike means that Augustine sees them all as people who are charged with maintaining peace and order. In fact, he tells the reader that it would be best for all citizens to maintain this mindset of humble self-sacrifice when he writes, "This doctrine, if it were obeyed, would be the salvation of the commonwealth."⁹⁹ To place this idea in a present-day context, Augustine would support the idea of the citizen-soldier who, whether he serves for a

---

⁹⁵ Ibid., 554.
⁹⁶ Ibid.
⁹⁷ Ibid., 486.
⁹⁸ Ibid.
⁹⁹ Ibid.

short period of time or for a lifetime, retains a desire for loyal service to his nation and to his fellow citizens.

Another benefit that the righteous ruler receives from creating an atmosphere where citizens and soldiers share the same values is that he is more likely to receive the support of his population during wartime. Furthermore, when the time comes to re-establish peaceful relations with the other nation, the people will be more likely to concur with the conditions of that peace. In a letter, Augustine writes, "And on this principle, if the commonwealth observes the precepts of the Christian religion, even its wars themselves will not be carried on without the benevolent design that, after the resisting nations have been conquered, provision may be more easily made for enjoying in peace the mutual bond of piety and justice."[100] The idea is that every member of the state, if ruled property, shares the same set of Christian values. This would allow the state to function properly and support its ruler's decisions. Of course, the idea of total war, in which the entire state is engaged in a conflict, would come much later. Still, even in Augustine's day, it would be accurate to say that wars supported by the populace have a greater chance of success and that any popularly supported peace agreement would have a greater chance of allowing a return to normal relations between the states.

A Treatise Concerning the Correction of the Donatists

What happens when the ruler does not have the required level of domestic tranquility? What should a ruler do when not all citizens share the same Christian values? In some ways, Augustine addresses these issues in his *Treatise Concerning the Correction of the Donatists*. In part, Augustine's treatment of the Donatists is outside of the scope of this work. After all, the concept of *jus post bellum* applies to the treatment of foreign states after a war has concluded. As discussed earlier in this study, the Donatists were perceived as a threat to internal stability. However, Augustine's views toward the issue of the Donatists say much about how he felt about the role of church-state relations. This point of view has an impact on what Augustine would say about the proper role of *jus post bellum* activities as they are applied in the present day. Also, given that the Donatists were persecuted during Augustine's lifetime, his views regarding how much violence was necessary in order to produce an end to conflict and a lasting peace can be described in practical, rather than theoretical, terms. To put this another way, unlike theoretical foreign wars, the Donatist controversy was an issue that had a profound, direct impact on Augustine's ministry. It would be interesting to

---

[100] Ibid., 485.

examine the lengths to which Augustine personally felt that the government should intervene in order to finally quell the issue.

During the Donatist controversy, Augustine's tolerance for imperial intervention grew greater as time passed. According to Charles Scalise, "During the early years of his controversy with the Donatists (c. 392-404) Augustine clearly advocates for a position opposing religious coercion."[101] At first, Augustine was open to debate and persuasion. As the controversy grew worse, though, he began to see the Donatists as a destabilizing force that tended to threaten the safety, and the souls, of other citizens within the Western Roman Empire. As time passed, he advocated for intervention to the point of "violent religious oppression."[102] The *Treatise Concerning the Correction of the Donatists* does much to explain his evolution over time.

Within the document, Augustine appears to find that violence is worthwhile if the end result is the preservation of souls. He writes, "Why, therefore, should not the Church use force in compelling her lost sons to return if the lost sons compelled others to their destruction? Although even men who have not been compelled, but only led astray, are received by their loving mother with more affection if they are recalled to her bosom through the enforcement of terrible but salutary laws."[103] Within this statement lies the implication that there is one and only one Christian church. Any other faith that calls itself Christian yet is not part of the Catholic Church is a form of heresy. Given that heretics, such as the Donatists, seek to take others away from the Catholic Church, violence against them is acceptable for their sake and for the sake of those they wish to convert to their faith.

In making this statement, Augustine implies that all salvation is found within the Catholic Church. It would be impossible for any pagan or for any Christian who belongs to a heretical sect to gain entrance into heaven. For the Donatists, their heresy is especially destructive because, in Augustine's opinion, their souls are at risk and their own ministry is at cross purposes with the ministry of the Catholic Church.

Another implication of this statement is that Augustine sees the Church and the state as being unified in purpose. The idea is that the Church

---

[101] Scalise, "Exegetical Warrants," 497.
[102] Ibid., 499.
[103] Augustine of Hippo, *A Treatise Concerning the Correction of the Donatists* in *A Select Library of the Nicene and Post-Nicene Fathers of the Christian Church*, Volume IV, *St. Augustine: The Writings Against the Manichaeans and the Donatists*, trans. J.R. King, ed Philip Schaff (Buffalo, NY: The Christian Literature Company, 1887), 642.

performs its normal ecclesiastical functions, such as the administration of the Sacraments, and spreads the gospel message, both inside and outside the boundaries of the state. That being said, it is the responsibility of the state to create the conditions that facilitate the Church's work. In the case of the Donatists, the Church was not structured in such a way that it could actively prevent a rival church from operating on a large scale. To put this another way, there was nothing that the Church could do to prevent the Donatists from congregating in their own churches, administering their own Sacraments, and spreading their own theology. Only the state had the administrative and military tools necessary to remove large-scale hindrances, such as these, to the Church's overall mission.

In making this argument, Augustine uses pastoral imagery, referring to the Church as the shepherd and believers as sheep who could be led astray. He writes, "Is it not a part of the care of the shepherd, when any sheep have left the flock, even though not violently forced away, but led astray by tender words and coaxing blandishments, to bring them back into the fold with the master when he has found them, by fear or even the pain of the whip, if they show symptoms of resistance."[104] Of course, he would have preferred to not need to use violence in order to bring these "lost sheep" back into the "fold." However, in his opinion, if violence is required in order to protect these Christians, then, for their own sake, it must be applied.

One counterargument that Augustine addresses in his *Treatise Concerning the Correction of the Donatists* is that the Church, which suffered under persecution from its founding until the time of Constantine, has no right to take advantage of its new position and persecute others. In addressing this issue, he is careful to make a distinction between vengeful persecution and loving correction. Augustine writes, "She persecutes in the spirit of love, they in the spirit of wrath; she that she may correct, they that they may overthrow; she that she may recall from error, they that they may drive headlong into error. Finally, she persecutes her enemies and arrests them, until they become weary in their vain opinions, so that they should make advance in the truth."[105] Unfortunately for Augustine, given the very circumstances that led to their rise, the Donatists were almost uniquely suited to suffer persecution and, in fact, remained relevant until the Arabs conquered the land several hundred years later.

In summary, *The Treatise Concerning the Correction of the Donatists* illustrates how much state-sanctioned violence Augustine could support. In

---

[104] Augustine, *Treatise*, 642.
[105] Ibid., 637.

alignment with his other writings, it is apparent that he truly did value peace and stability, but there was an ever-present theological aspect to his views on conflict. In short, Augustine saw spiritual fulfillment as the highest good and viewed the state and the Church as partners in a holistic effort to maintain, support, and spread orthodox Christianity.

To summarize the chapter, Augustine's views on war were very much influenced by his cultural context. The Roman Empire had very recently experienced both rapid Christianization and rapid decline. This made it easy to draw the conclusion that one had caused the other and, to his credit, Augustine addressed this issue successfully by arguing that Rome had significant issues since well before the time of Jesus. Furthermore, he discussed what was, in his opinion, the new relationship between state and church as well as with ruler and citizen now that the Roman Empire was not only a Christian state but also the sole Christian state in the world.

In brief, Augustine believed that the ruler has a responsibility to create conditions of lasting peace, both within the state and outside its boundaries for the purposes of aiding humanity in its goal to form a closer relationship with God. To put this another way, it is much easier to minister to the spiritual needs of a relatively safe and secure people. Disciplined and carefully considered violence may be necessary in furtherance of this goal, but it is understood that this level of violence would exist to prevent an even higher level of suffering. Furthermore, both the state and the citizen are unified in this effort, with the ruler serving as God's appointed leader and the citizen serving in a supporting role.

This study has demonstrated that Augustine's ideas have been processed and carried forward into the present day. However, it is apparent that his ideas regarding just war were formulated within a context that no longer exists. Western society has secularized, and models of government have changed. How much have these principles been altered and what form did they finally take by the beginning of the twentieth century? The next chapter will discuss that evolution.

3. JUST WAR DEVELOPMENT AND REFINEMENT

Post Augustinian Development

Between the fifth century and the twentieth century, Augustinian thought, as pertaining to just war, was refined and codified. It was Thomas Aquinas who formulated the just war theory categories that are so recognizable in the present day.[106] However, other just war theory thinkers, such as Gratian, Luther, Grotius, and Kant were instrumental in making significant contributions to the field in response to a changing world. The issue was that, since the time of Augustine, the very nature of Western government was changing. As the Middle Ages gave way to the Enlightenment, the idea of the "Christian" ruler and of a "Christian" government also changed.[107] Yet, the basis of just war theory remained. These thinkers gradually took just war theory out of its religious context and placed it into a moral context without changing the core tenets of Augustinian thought, such as the need to go to war reluctantly, the requirement to keep violence to the minimal possible level, and the necessity of ensuring that the war has the end goal of preventing a greater amount of suffering. Interestingly, though, with the exception of Kant, these thinkers tended to focus on conduct before and during a conflict.

---

[106] Gregory M. Reichberg, "Thomas Aquinas on Military Prudence," *Journal of Military Ethics* 9, no. 3 (2010): 262.
[107] G.R. Evans, *The Roots of the Reformation: Tradition, Emergence, and Rupture*. 2nd ed. (Downers Grove, IL: IVP Academic, 2012), 379.

Gratian was, perhaps, one of the more notable just war thinkers between the eras of Augustine and Aquinas.[108] Briefly, he was a twelfth-century Benedictine monk who compiled canon law into the *Decretum Gratiani*. This work was important because it remained the standard for Church law for the next several hundred years. Gratian included the principles of just war theory, such as legitimate authority and righteous intent, as derived from Augustine, within his text.

Gratian found that the waging of a just war was more of a matter of responding to injustice rather than defending one's own territory.[109] The idea was that states have a moral responsibility to prevent aggression and stamp out tyranny wherever it may be. Naturally, as with Augustinian thought, this would have included internal threats. The basic principles of legitimate authority, proper reasons to go to war, and rightful intention, as noted within Augustinian thought, are found within Gratian's compiled work. However, there is not as much emphasis on the ruler being a God-ordained "Christian" ruler even though the *Decretum* was written for Western Christian rulers. As Rory Cox writes, "Gratian's interpretation of proper authority was that it resides in any person or institution that could issue an authoritative edict."[110] Other scholars interpret this as meaning that Gratian placed secular authority in the hands of both civil and Church leaders.[111] For Gratian, it was the office of rulership that conveyed the authority to rule as well as the responsibility to rule justly.

For this study, it is important to emphasize that, while Gratian was, in fact, compiling Church law, he did not quite add the spiritual dimension to just war theory that was so intrinsic to Augustinian thought. As discussed, Augustine viewed war as a means to reduce injustice for the purpose of allowing people to focus on spiritual matters rather than their own temporal needs. In this way, the state was to provide the breathing room that the church required in order to fulfill its function. Gratian made just war more of a moral issue as he attempted to incorporate Greek and Roman thought, in addition to Augustinian thought, into a code of laws that were intended

---

[108] Rory Cox, "Gratian (Circa 12th Century)" in *Just War Thinkers: From Cicero to the 21st Century*, eds. Daniel R. Brunstetter and Cian O'Driscoll (London, UK: Routledge, 2017), 34.
[109] Cox, "Gratian," 44.
[110] Ibid., 40.
[111] Melodie H. Eichbauer, "The Bishop with Two Hats: Reconciling Episcopal and Military Obligations in Causa 23 of Gratian's *Decretum*" in *Civilians and Warfare in World History*, eds. Nicola Foote and Nadya Williams (New York, NY: Routledge, 2018), 122.

for practical use.¹¹² In a way, he began to move just war theory away from its strictly Christian roots, though he still advocated for wars against non-Christians for the purpose of forcible conversion.¹¹³ Later, Thomas Aquinas would take just war theory one step further into secular thought in his *Summa Theologica*.

Thomas Aquinas wrote the *Summa Theologica* in the thirteenth century in order to summarize the theology of the Catholic Church. Within this work, Aquinas codified the just war principles that have been discussed throughout this study. In brief, he found that, for wars to be justified, the nation entering into war must have a sovereign authority, a just cause, and a rightful intention. Furthermore, within the confines of war, violence must be proportional, must be kept to a necessary minimum, and must exclude non-combatants.¹¹⁴ None of these principles are contrary to Augustinian thought, but it is important that Augustine's thoughts on war are actually codified within Aquinas's work.

Aquinas sought to finally declare, with ecclesiastical approval, the parameters of a just war. According to Gregory Reichberg, "When Thomas Aquinas wrote of just war in the thirteenth century, he clearly did not seek to innovate, but was intent rather on establishing a place for it on his map of the moral universe."¹¹⁵ Reichberg's point about Aquinas's use of a moral framework is salient and connects it to Gratian's efforts.

Though the *Summa Theologica* was an ecclesiastical product and, essentially appropriated Augustine's thoughts on war, it was written within a different context. First, Aquinas came from a military family and held a "keen interest in the profession of arms."¹¹⁶ Interestingly, though, while Gratian did speak of using armed force for the purpose of conducting wars against internal and external unbelievers, Aquinas tended to advocate for refraining from religious wars unless they were necessary.¹¹⁷ The study of war and the moral questions surrounding conflict were important to him. Second, the *Summa Theologica* had much to do with matters of secular statehood. As Reichberg states when he discusses the right of the state to wage secular war

---

¹¹² Cox, "Gratian," 39.
¹¹³ Gregory M. Reichberg, *Thomas Aquinas on War and Peace* (New York, NY: Cambridge University Press, 2017), 261.
¹¹⁴ R.W. Dyson, trans., *Aquinas: Political Writings* (Cambridge, UK: Cambridge University Press, 2002), 239-242.
¹¹⁵ Reichberg, *Thomas Aquinas*, 257.
¹¹⁶ Ibid., 12.
¹¹⁷ Ibid., 263.

in pursuit of national interest, "The defense of faith no longer occupies the central stage. Instead, the resort to force is justified by reference to the well-being of the *republica*."[118] This slow secularization and this movement away from the idea of a Christian emperor or king as the source of authority will be important as the development of just war theory continues into Luther's day.

Also of note, Aquinas' vows as a Dominican likely played a role in his views on war. In his mind, obedience to higher authority was necessary but not absolute. As Reichberg states, "For Aquinas obedience – whether for persons under (1) religious vows, (2) civil authority, or (3) military command – can never bind absolutely, such that it unqualifiedly overrides all other concerns."[119] Essentially, this means that, in war, agents of the state are required to use their own judgment when deciding how to execute their leaders' commands. This concept works well with the *jus in bello* ideas derived from the Augustinian model though, of particular note, it would be crucial that agents of the state become virtuous themselves in order to exercise this level of discernment.[120]

Just war theory made a further movement towards secular applicability during the Reformation. Augustine, Gratian, and Aquinas all wrote within a Catholic context even though their thoughts applied to the ability of secular leaders to commence and wage war. In their minds, the spiritual component of war was ever-present even as the conduct of war slowly moved from a strictly Christian to a generally moral issue, as argued within this work. Still, it was during the time of Luther that the idea of Christian warfare was completely divorced from the papacy.

Luther wrote extensively on war but, perhaps, his most valuable contribution to just war theory came from *Temporal Authority* in which he argued from the New Testament to claim that Christians have a duty to obey secular authorities. He writes, "The Christian submits most willingly to the rule of the sword, pays his taxes, honors those in authority, serves, helps, and does all he can to assist the governing authority."[121] Luther's context was that of secular German princes who did not derive their right of rulership from the papacy and based on this, he found that war was completely a matter for

---

[118] Ibid., 7.
[119] Ibid., 239.
[120] Ibid., 239.
[121] Martin Luther, "Temporal Authority: To What Extent It Should Be Obeyed" in *Martin Luther's Basic Theological Writings*, 3rd ed., eds. Timothy F. Lull and William R. Russell (Minneapolis, MN: Fortress Press, 2012), 435.

Just War Development and Refinement

the secular ruler. As James Johnson writes, "Advice from others should be solicited, but in the end, the sovereign's right to use force comes from his responsibility for order, justice, and peace."[122] This is not to say that secular authorities since the time of Augustine had not been acting on their own authority. However, up until this point, Western secular rulers had derived much of their authority from the papacy. After the Reformation, the papacy had much less influence over the sovereign and, from this point on, secular authority was often seen to be superior to sacred authority.[123] Still, the Christian roots of Augustinian just war theory remained and were syncretized with moral and natural law.

Hugo Grotius, who wrote extensively on just war theory after the time of Luther, essentially argued that there was no gap between a Christian understanding of just war theory and a secular, moral understanding of the theory's precepts.[124] He examined the law of warfare from the perspective of natural law. As Anthony Lang writes, "Grotius' contribution to the natural law tradition was to explore morality in the context of the least law-bound human practices, war."[125] Have there been inherent truths to the conduct of war that had always existed throughout human history? According to Lang, Grotius answered in the affirmative. He writes, "Grotius argued that there was no clash between natural law and the beliefs of a Christian."[126] This did not mean that all societies had the same exact rules but, rather, that many ideals of justice in war were common and fit well within a Christian moral framework.

This is an interesting inflection point because it means that Augustine's writings retained their relevance up until this period even though just war theory had become more of a secular construct. Of course, Grotius' examination of natural law theoretically applied to all cultures at all times. This study is more narrowly focused on just war theory in a Western context. Still, a possible implication of Grotius' work is that Augustinian just war theory is in alignment with God's moral code even though it has become, in practice, a secular body of thought. Of interest though, some of the reasons

---

[122] James Turner Johnson, "Aquinas and Luther on War and Peace: Sovereign Authority and the Use of Armed Force," *The Journal of Religious Ethics* 31, no. 1 (Spring 2003): 17.
[123] Evans, *Roots of the Reformation*, 388-389.
[124] Hugo Grotius, *The Rights of War and Peace*, ed. Knud Haakonssen (Indianapolis, IN: Liberty Fund, 2005), 1762.
[125] Anthony F. Lang, Jr., "Hugo Grotius (1583-1645)" in *Just War Thinkers: From Cicero to the 21st Century*, eds. Daniel R. Brunstetter and Cian O'Driscoll (London, UK: Routledge, 2017), 134.
[126] Ibid., 135.

why Augustine penned his thoughts on war remain absent from the conversation. Referring back to Augustine's thoughts on war, the idea of the Christian ruler giving his subjects a safe space to become better Christians has, in some ways, passed away. Still, rules requiring decent conduct during war and sound reasons for entering into war remain.

In all of the time, though, between Augustine and Grotius, it is fascinating that no major just war thinker had proposed a framework for conduct after the conclusion of a war. Brian Orend emphasizes this point in his discussion of Kant when he writes, "One of the most creative things about Kant's reflections on warfare deals with peace treaties, or the proper endings of wars. This is totally unlike thinkers both before and after him, obsessed as they have been with the outbreak of wars."[127] What made Kant the first *jus post bellum* thinker and why had no one seriously considered the concept before him? The answer lies in Kant's definition of "peace."

During the Enlightenment, absolutist forms of government were giving way to republics that were, theoretically, governed by the people. For Kant, peace, not war, was the natural state of affairs between republics as the desires for personal freedom and for free trade would necessarily cause citizens to promote peace.[128] If, by chance, war commenced due to the actions of an aggressor, then it was critical that the belligerent nation be transformed upon the conclusion of hostilities so that it would be unable to commit such actions again.[129]

In some ways, Kant's ideas foreshadowed the lessons learned at the conclusion of World War I but, still, it may be reasonable to argue that his view of society was, perhaps, too optimistic. Just war scholars note that he has an almost equivocal approach to war, vacillating between realism and pacifism but, in the balance, he appears to see war as an occasional necessity.[130] His thoughts are important to the arguments made within this work because he has continued the trend of moving just war theory further away from many of its Augustinian roots. Furthermore, his idea of a lasting peace comes from a place of fully secular, rather than Christian, thought.

---

[127] Brian Orend, "Immanuel Kant (1724-1804)" in *Just War Thinkers: From Cicero to the 21st Century*, eds. Daniel R. Brunstetter and Cian O'Driscoll (London, UK: Routledge, 2017), 174.
[128] Ibid.
[129] Ibid., 175.
[130] Howard Williams, *Kant and the End of War: A Critique of Just War Theory* (New York, NY: Palgrave Macmillan, 2012), 41.

Francis Lieber also deeply considered the means by which to bring about a lasting peace. His goals, though, differed from Kant's in significant ways. Lieber sought to bring about the end of war, and thus establish peace, by making war end as quickly as possible. Stephanie Carvin writes, "The Lieber Code makes it clear that war is a means to an end, not an end in and of itself, and therefore limited. The object of war is not killing, but to achieve a political end, normally the return to peace."[131] The idea was that the war itself would be so decisive that a lasting peace would be the expected outcome. In order to properly produce this outcome, he put tight restraints around soldiers' conduct in war by generally disallowing immoral conduct while allowing overwhelming action against military targets.

His original set of documents, known as the Lieber Code, was written during the American Civil War and informed the modern-day Geneva Convention.[132] Within the Lieber Code, the *jus in bello* actions discussed thus far were included in addition to proscriptions against unnecessary suffering that would seem familiar to modern ears. As Carvin writes, "it is possible to find in the Lieber Code many of the provisions that would make themselves into future national and international documents on the laws of war, including bans on slavery (Articles 41–44), not giving quarter (Article 60), poison (Article 70), unnecessary suffering (Article 71), and torture (Article 16)."[133] However, his views on peace were based on the usage of intense military action, meaning that he had no *jus post bellum* doctrine as would be formulated in later times.

In summary, between the time of Augustine and the dawn of the early twentieth century, many of Augustine's ideals had remained within just war theory even though much of the context had changed. Augustine envisioned a Christian leader wisely handling secular matters in such a manner that the citizens were largely free to pursue spiritual matters. This ruler would engage in war reluctantly and only if necessary to correct an injustice or preserve the well-being of his people. By the time of the early twentieth century, the idea of righteous secular leadership, supported by a conscientious citizenry, remained, but the spiritual dimension of Augustinian just war theory had been stripped away. That being said, during the twentieth

---

[131] Stephanie Carvin, "Francis Lieber (1798-1872)" in *Just War Thinkers: From Cicero to the 21st Century*, eds. Daniel R. Brunstetter and Cian O'Driscoll (London, UK: Routledge, 2017), 187.

[132] Adam Roberts, "Foundational Myths in the Laws of War: The 1863 Lieber Code, and the 1864 'Geneva Convention'," *Melbourne Journal of International Law* 20, no. 1 (July 2019): 159.

[133] Carvin, "Lieber," 188.

century, the idea of *jus post bellum* will be added to the just war paradigm and, importantly, it will be retroactively attributed to Augustine.

Turning Points: The World Wars

By the commencement of World War I, a robust *jus ad bellum* and *jus in bello* framework was in place that, though developed over time, still maintained its base Augustinian principles. With the war, though, many of these principles were broken and, most importantly, the need for a strong *jus post bellum* structure became apparent.

The standard *jus ad bellum* principles that should have been applied at the commencement of World War I were not followed.[134] The European powers were led by legitimate sovereigns, but the war was not characterized by either a just cause or a rightful intention. Rather, beginning with the assassination of Archduke Franz Ferdinand of Austria, a complex series of alliances drew the nations of Europe into a conflict that lasted four years and stretched their resources to the breaking point.

As the war became a long-term war of attrition, the typical *jus in bello* principles that had been well-recognized were eventually disregarded. The war was known for the implementation of unrestricted submarine warfare, and the commitment of genocide.[135] Though the war began with much more stringent observation of the ideals of *jus in bello*, desperation eventually gave way to expediency. A long-term blockade, for instance, was used as an attempt to break the stalemate on the Western Front and unrestricted submarine warfare became a desperate act to push Britain out of the war.[136] To frame this dynamic in an Augustinian context, The First World war could not have been considered a just war because key principles such as prohibitions against harming civilians and minimizing suffering were slowly discarded.

For the purposes of *jus post bellum* study, perhaps the most glaring example of a failure to establish the Augustinian ideal of a just peace was the Treaty of Versailles. When the armistice was finally signed, Germany was forced to sign a document that could be viewed as having been overly punitive. As Iasiello writes, "It directed Germany to give up some of its most valuable territories, place the Rhineland under an allied protectorate for

---
[134] Nicholas Fotion, *War and Ethics: A New Just War Theory* (New York, NY: Continuum, 2007), 57-59.
[135] Ibid, 64.
[136] Ibid.

fifteen years, and bear both occupation costs and painful postwar reparations."[137] Considering that, by the end of the war, German armies remained intact and Germany itself had not been conquered, it is easy to understand why many Germans felt that their political class had betrayed them.

Later, German resentment coupled with a worldwide economic downturn created the conditions by which the nation could fall under the sway of a militaristic, totalitarian regime. The problem was not confined to Germany, however. Both Italy and Japan, though they were victors in the First World War, also felt that they were treated unfairly by the Allied powers.[138] These nations, too, would come under the sway of fascist governmental structures. In many ways, the very peace that ended one war caused another. As Iasiello writes, "This absence of postwar vision negated, for all practical purposes, any hope of a just and lasting peace. Some would blame Europe's subsequent economic chaos and wounded nationalism, the birth of totalitarianism, and ultimately World War II itself on this lack of war-termination vision."[139] Fortunately, the peace treaties signed at the conclusion of World War II, and the *jus post bellum* activities that began afterward, were much more successful at creating the conditions that should have led to a lasting peace.

Europe and Asia were devastated at the conclusion of World War II, but it is apparent that some of the lessons of post-World War I were learned. As Carsten Stahn writes, "The peace settlements after World War II present a slightly more nuanced picture. Human rights clauses and provisions for criminal adjudication became integral features of peace treaties with former enemy powers."[140] The issue of criminal adjudication was notable. At the end of World War I, nations were held accountable by the Allied powers, but individuals emerged relatively unscathed. In Germany and Japan after the Second World War, however, tribunals set out to try belligerent leaders accused of war crimes within the context of a comprehensive legal framework. From an Augustinian perspective, this was a step in the proper

---

[137] Iasiello, "Jus Post Bellum," 38.
[138] Michael S. Neiberg, *The Treaty of Versailles: A Very Short Introduction* (Oxford, UK: Oxford University Press, 2018), 61.
[139] Iasiello, "Jus Post Bellum," 38.
[140] Carsten Stahn, "*Jus Post Bellum*: Mapping the Discipline(s)," *American University International Law Review* 23, no. 2 (2007): 318.

direction as individual accountability is considered a method of correcting injustice.[141]

The Allied powers also made tremendous efforts to ensure that the defeated nations, as well as other affected nations, were properly rebuilt. As Stahn writes, "In the cases of Germany and Japan, victory was combined with economic, social, and legal reconstructions."[142] In Europe, the Marshall Plan was the means by which the United States paid for economic reconstruction, humanitarian aid, and political reinforcement. A major goal of the plan was to keep Western Europe outside of Soviet influence, but the idea of producing a safe, lasting peace remained the overarching goal.[143] Similar efforts were performed in Japan. General MacArthur served as the military governor of Japan for six years. During this time, he instituted democratic political reforms and economic reconstruction plans that were notably successful given that Japan had experienced large-scale fire-bombings and two atomic attacks.[144] By the early 1950s, Japan appeared to be a stable, well-functioning Western-style state.

In summary, the first half of the twentieth century saw the emergence of *jus post bellum* activities that appeared to be in alignment with Augustinian thought. Personal accountability, rebuilding efforts, and direct aid to alleviate suffering were all practices that had the intention of ensuring stability within war-torn regions. Theologically, the spiritual aspect of these *jus post bellum* activities is noticeably absent. Florian Demont-Biaggi describes this dynamic as a "secular version" of Augustinian thought in which "military success is all about establishing a well-ordered concord, a state of peace."[145] Still, the renewed ability of the citizenry to pursue their life goals can be considered a positive development by Augustinian standards. With the emergence of *jus post bellum* standards after the Second World War, the concept faced continual refinement until the present day.

Jus Post Bellum Refinement

---

[141] Mattox, *Saint Augustine*, 124.
[142] Stahn, "*Jus Post Bellum*," 318-319.
[143] Ibid., 319-320.
[144] Francis Pike, *Hirohito's War: The Pacific War, 1941-1945* (London, UK: Bloomsbury Publishing, 2016), 1061-1068.
[145] Florian Demont-Biaggi, "Causation, Luck, and Restraint in War," in *Jus Post Bellum: Restraint, Stabilisation, and Peace*, ed. Patrick Mileham (Boston, MA: Brill, 2020), 42.

The concepts that would come to be crucial aspects of a *jus post bellum* framework truly came into being during the middle of the twentieth century. With the ending of two worldwide conflicts, the commencement of the Cold War, and the initiation of wars in the Middle East, ideas regarding accountability, justice, and reconstruction were put into action.[146] It was found that, while the key concepts of a *jus post bellum* framework could produce the desired lasting peace, each pillar had its own limitations and potential pitfalls.

War Crimes and Accountability

Though *jus post bellum* is a relatively new concept, the idea of holding belligerents, both collectively and personally, accountable for their actions is long-standing. It has been demonstrated that justice needs to be a component of a peace settlement. Eric Patterson makes an excellent point when he compares the results of the two treaties of Paris that concluded the Napoleonic Wars and finds that it was only the final treaty, the one that held Napoleon Bonaparte personally accountable for his actions, that prevented the resurgence of hostilities.[147] After his forced abdication in 1814, Napoleon retired to a comfortable existence on the nearby island of Elba, but later, after a return to power and the Battle of Waterloo, Napoleon received much harsher peace terms and was required to face exile on the distant island of St. Helena.[148] Following this decisive conclusion to Napoleon's career, Europe remained relatively peaceful for the next century.

Still, post-war justice is a complicated, multifaceted topic. What does it mean to distribute equal justice after a war? In answer to this question, Patterson writes, "Justice is incurring what one deserves. It is getting one's just deserts. In a post-conflict setting this means that those who had responsibility for political and military choices that violated basic principles of humanity and/or the laws of armed conflict are accountable for their decisions or their deeds."[149] Ideally, justice provides some form of restitution to those who were harmed and it may even help serve as a deterrent to others who would choose to harm others.[150] It is also possible that, as with the case

---

[146] Iasiello, *Jus Post Bellum*, 39.
[147] Patterson, *Ending Wars Well*, 68.
[148] J. David Markham, *The Road to St. Helena: Napoleon after Waterloo* (South Yorkshire, UK: Pen and Sword, 2008), 132.
[149] Patterson, *Ending Wars Well*, 69.
[150] Brian Orend, "*Jus Post Bellum*: The Perspective of a Just-War Theorist." *Leiden Journal of International Law* 20 (2007): 580.

of Napoleon, justice provides a civilized means of removing a ruler from the political stage.

As an aspect of *jus post bellum* justice, personal accountability can be applied to both political leaders and to the military members who act as their agents. As Patterson notes, "Efforts at Justice, such as holding individual soldiers accountable for their misdeeds or political elites accountable for the orders they gave, reinforce the moral order."[151] It is a hallmark of twentieth-century post-war activities that leaders faced criminal prosecution for their actions.[152] For the purpose of this conversation, notable war crimes include actions that would have fallen clearly outside of internationally accepted rules of war and the *jus in bello* framework, such as intentional harm to civilians, torture of prisoners, and unwarranted property damage.

The International Military Tribunal at Nuremberg, which was held from the end of 1945 to the end of 1946, tried over twenty German officials for war crimes within the context of an international court of law.[153] They were charged with war crimes, evidence was presented against them, they were allowed to deliver an adequate defense and, in the end, those who were convicted of war crimes received sentences ranging from imprisonment to execution. Similar trials took place in Tokyo during the International Military Tribunal for the Far East with similar results.[154]

It is important for the sake of *jus post bellum* development that these trials were the first of their kind. Though Germany had been held responsible, as a nation, for the First World War and Napoleon had been held responsible for the wars of his era, accountability, in those cases, came through peace treaties that were forced by the victors upon the defeated. In the trials after the Second World War, the belligerents were able to exercise their rights, as evidenced by the fact that some defended themselves successfully.

Justice in a *jus post bellum* context also involves an element of restitution. It is not enough that those who commit war crimes face criminal charges for their actions. There must be some effort to make the victims of war whole again. As Patterson writes, "Justice also provides something for

---

[151] Patterson, *Ending Wars Well*, 70.
[152] Bass, "Jus Post Bellum," 406.
[153] Devin O. Pendas, *Democracy, Nazi Trials, and Transitional Justice in Germany, 1945-1950* (Cambridge, UK: Cambridge University Press, 2020), 37.
[154] Yuma Totami, *The Tokyo War Crimes Trial: The Pursuit of Justice in the Wake of WW II*, 1st ed. (Boston, MA: Harvard University Asia Center, 2008), 1.

the victims. An element of Justice for some of those who were wronged is vengeance – the vindication of their righteous indignation, suffering, and loss."[155] Restitution both acknowledges the loss and makes a good-faith effort to help the affected persons recover from the conflict.

The idea of restitution is in alignment with Augustinian ideals. Patterson notes, "The idea that Justice has retributive features goes back to the earliest era of the just war tradition to Augustine, who argued that punishing those who provoke war is morally appropriate."[156] Accountability promotes lasting peace because it acknowledges wrongdoing, places consequences on those who commit unlawful actions, provides restoration, and discourages further unlawful action.

The exact nature of post-war restitution can be decided after the conflict has ended and, historically, has tended to take place on a national level. As Patterson notes, "Restitution may take many forms, and the financial commitment may, in fact, be a 'punishing' burden on a government, as Germany experienced at the end of the First World War when it had 'accepted' war guilt, sacrificed (or restored) land to its neighbors, ceded its navy and much of its merchant marine to the victors, and paid reparations to the Allies."[157] In some ways, it makes sense to have nations account for the war crimes of their leaders. Given the amount of destruction caused by war, rebuilding homes, returning land, and providing proof that the belligerent will not take hostile action again requires national-level sacrifice.

Though the process of providing post-war justice has had notable successes, it has also demonstrated that it has significant limitations. First, it is possible that, when justice is improperly applied, it can undermine the peace that it seeks to achieve. As Patterson notes, "The pursuit of Justice – its implementation in law and policy- should not erode fragile post-conflict Order. It should not undermine the ability of belligerents to negotiate in good faith nor should it be so vengeful as to sow the seeds of future war."[158] The failure of the peace of Versailles to prevent World War II is, perhaps, the most noteworthy example, but more recent events have also been less than successful. The capture and prosecution of Saddam Hussein during the American conflict in Iraq, for example, did little to quell the violence there.

---

[155] Patterson, *Ending Wars Well*, 71.
[156] Ibid., 75.
[157] Ibid., 73.
[158] Ibid., 69-70.

Providing justice in war-torn, failed states has additional problems. Deterrence is useful, but only in a unified, organized state. Patterson writes, "Justice-as-deterrence has a weak track record in international affairs, especially in contexts of weak institutions, fragile states, and especially stateless spaces like Yemen, Somalia, and parts of Colombia."[159] A necessary prerequisite to criminal prosecution is the ability of a government entity to apprehend the offender. In some instances, such as when it is only possible to negotiate with the accused but not to bring the person to justice, this can be a difficult process.[160]

In summary, though the idea of justice in a *jus post bellum* environment has changed dramatically since World War I, significant challenges remain. There have been great strides in efforts to punish wrongdoers and provide justice for victims even if the ideal of perfect justice remains elusive. As Patterson notes, "Elite punishment, or punishment of a subsection of aggressors, is better than no justice at all, and its limits testify to the real-world boundaries of international life, and perhaps to the wise restraint of those who could seek more but choose not to."[161] Put another way, the world is a complicated place and, though righteousness in all cases may be desirable, justice should take real-world realities into account.

Still, it can be argued that *jus post bellum* justice stands upon two pillars: personal accountability and national accountability. Personal accountability involves holding individuals criminally responsible for acts that breach long-standing laws of armed conflict. Accused war criminals are given the right to a fair defense. Still, there are two notable roadblocks to individual justice. First, the accused must be located and apprehended. This could be a problem within a failed state. Second, it is possible that leaders who are in charge of a failing military effort may choose to continue to fight, thus causing more suffering, rather than surrender because they simply have little to lose.[162]

National accountability also has value, if properly applied. However, it, too, has its pitfalls. The lessons of post-World War One remain. A nation that has been required to give away too much in terms of national honor, territory, and financial reparations may become resentful, thus sowing the seeds of a future conflict. However, despite these obstacles, rational attempts

---

[159] Ibid., 71.
[160] Bass, "Jus Post Bellum," 404.
[161] Patterson, *Ending Wars Well*, 77.
[162] Bass, "Jus Post Bellum," 405.

at providing justice have proved better than no efforts at all.[163] To place this argument back into an Augustinian context, a concentrated desire to create a lasting peace through justice has proved to be a better option than a disregard for postwar justice. Theologically, while the perfect justice imagined in the City of God may never be possible in this lifetime, nations have a duty to attempt to provide the fullest extent of justice possible.

Economic Reconstruction

Economic reconstruction became something of a twentieth-century postwar phenomenon because, frankly, before that time, very few nations had the capacity to accomplish it. Shortly after World War Two ended, in what became known as the Truman Doctrine, the United States took the lead in ensuring that Greece and Turkey received economic assistance.[164] This was not simply about rebuilding after a war as Turkey had very little involvement in the conflict. The main focus was ensuring that their governments did not become Soviet-aligned communist regimes. This decision had far-reaching implications for economic reconstruction efforts afterward because the United States made a commitment to ensure the stability of friendly foreign governments.

A similar calculus took place shortly afterward in what became known as the Marshall Plan. The European Recovery Program provided approximately $13 billion in aid to the nations of western and southern Europe over a four-year period and the focus of the plan was to provide the necessary support that allowed these nations to energize their industrial and agricultural production[165]. The plan also included elements that enabled financial stability and free trade. Again though, while this era of reconstruction had noble intentions, the political goal of subverting Communist expansion remained.

There was a similar rebuilding project that took place in Japan, though it was under the rule of a military governorship.[166] Notably for this study, Japanese industry remained intact, though it had previously been

---

[163] Patterson, *Ending Wars Well*, 77.
[164] Athanasios Lykogiannis, *Britain and the Greek Economic Crisis, 1944-1947: From Liberation to the Truman Doctrine* (Columbia, MO: University of Missouri Press, 2002), 223-224.
[165] Michael Holm, *The Marshall Plan: A New Deal for Europe* (New York, NY: Routledge, 2017), xv.
[166] Dayna J. Barnes, *Architects of Occupation: American Experts and Planning for Postwar Japan* (Ithaca, NY: Cornell University Press, 2017), 163.

consolidated into only a few companies. Therefore, the United States reorganized Japanese industry in the interest of promoting a free-market economy, and land reform was needed for similar reasons.[167] Furthermore, since Japan historically had difficulties with acquiring raw materials, the United States made efforts to provide Japan with the raw materials it needed to thrive.[168] The political calculus was also somewhat different. The issue was less about Communist expansion and more about creating a stable partner.

These *jus post bellum* economic reconstruction efforts presented a picture of what similar efforts would look like until the present day. First, it is apparent that only a victor who had not started an unrighteous war but, rather, was responding to aggression could possibly become involved in economic reconstruction. James Pattison writes, "There should be a presumption against belligerents rebuilding. It seems that in some cases (although not all) they will lack the right to rebuild and, even if they do have the right, other agents may be in a better position to rebuild."[169] The defeated belligerent nation may simply not have the resources to rebuild. This was certainly the case with Germany and Japan. If the belligerent nation is victorious or if the involved nations are mutually exhausted by the conflict, economic reconstruction may still be unlikely or impossible. To use a previous historical example, American reconstruction efforts under the Truman Doctrine began because Great Britain had withdrawn its assistance in Greece.

A second principle of *jus post bellum* reconstruction has been that the moral responsibility for rebuilding falls upon the nation that is able to fund the effort. Pattison uses the example of a mutually exhaustive war in Central America as an example of a case where the United States should rebuild even if it had no role in the conflict.[170] The cost to the United States would be reasonable given the nation's wealth. Furthermore, by not assisting, the United States would be condemning the region to further unrest. It is true that the belligerent in this conflict should be responsible for reconstruction. Practically, though, there would be no means to rebuild these broken economies without foreign intervention.

A third principle of *jus post bellum* reconstruction is that the nations that were involved in the conflict may need outside assistance because their

---

[167] Barnes, *Architects of Occupation*, 48.
[168] Ibid.
[169] James Pattison, "Jus Post Bellum and the Responsibility to Rebuild," *British Journal of Political Science* 45, no. 3 (2015): 658.
[170] Ibid., 655.

societal structures and norms would simply not allow economic revitalization even if the money were available. As Pattison writes, "Those that are culpable for injustice may possess reparative duties and perhaps should pay for the rebuild at least fundamentally, but this does not mean that they should rebuild."[171] To use Japan as an example, even if it had the funding to rebuild its industry, returning to a system where a few companies dominated their markets meant that these companies would have hindered economic growth in the same way that cartels hinder development in other countries. For the good of the nation, these conglomerates needed to be liquidated, but it is doubtful that this could have taken place without external involvement.

Though the programs discussed thus far appeared to have been successful, economic reconstruction efforts within other nations since that time have experienced mixed success. After the Second Gulf War, which began in 2003, one American objective was to rebuild Iraq's infrastructure. However, the project seemed to have been hindered from its commencement. For example, as Darrell Cole notes, "The U.S. gave contracts only to companies politically connected to the U.S."[172] As a result, the Iraqis never truly gained ownership over these projects and, instead, saw them as foreign intervention. One impact was that the nation recovered to an extent, but the long-term damage remained. As stated earlier, it is generally best if the reconstruction effort is done by a foreign entity. However, an argument could be made that the United States should have turned over control of Iraqi reconstruction to trusted agents in Iraq much earlier.

In Afghanistan, two decades of economic reconstruction ended in failure. Even before American involvement in the country, Afghanistan had been engaged in international conflict for several decades and its economy was brittle, at best. While American efforts bore some fruit, permanent widespread efforts at economic reconstruction remained elusive.[173] Rouven Steeves, who writes extensively on the Afghan reconstruction effort, acknowledges the immense challenges posed by rebuilding a war-torn distant nation, but he claims that much of the blame for the failure of reconstruction in Afghanistan lies in the fact that the Taliban was never truly defeated.[174]

---

[171] Ibid., 655.
[172] Darrell Cole, "The First and Second Gulf Wars" in *America and the Just War Tradition: A History of U.S. Conflicts*, eds. Mark David Hall and J. Darryl Charles (Notre Dame, IN: University of Notre Dame Press, 2019), 265.
[173] Orend, *"Ju Post Bellum,"* 586.
[174] Rouven Steeves, "The War on Terror and Afghanistan" in *America and the Just War Tradition: A History of U.S. Conflicts*, eds. Mark David Hall and J. Darryl Charles (Notre Dame, IN: University of Notre Dame Press, 2019), 285.

Steeves has a salient point as it is exceedingly difficult to rebuild a nation in the face of an ongoing, active insurgency, though it is also important to note that this rebuilding effort was likely hampered by an incomplete understanding of Pashtun culture.

In summary, the history of *jus post bellum* reconstruction efforts since the twentieth century has been one of trial and error. After World War One, it was apparent that greater efforts needed to be made in order to ensure a lasting peace. After World War II, efforts at economic reconstruction were largely successful. Even if the motives of the victor were not entirely altruistic, it is undeniable that the economic structures in Europe and Japan recovered relatively quickly. However, later history has proven that economic reconstruction efforts, while never inexpensive or simple, could fail if improperly initiated.

Economic reconstruction has earned its place with the *jus post bellum* framework, and it has demonstrated that it can help produce a lasting peace that would be in alignment with Augustinian thought. To place reconstruction in an Augustinian context, the Roman Empire likely did not have the resources to lead a nationwide rebuilding effort. That being said, if Rome had been able to rebuild war-torn nations, Augustine would likely have approved of the efforts. Mattox essentially makes this argument when he finds that, in a world with one national superpower, that nation is charged with preserving the peace.[175] Notable present-day just war theorist Michael Walzer makes a similar argument when he finds that, despite the cost, long-term intervention can be worthwhile.[176] From an Augustinian theological perspective, a prosperous Christian government would have an obligation to assist the Christians of other nations whose economic situation hinders fellowship and spiritual development.

However, economic reconstruction is only one of many *jus post bellum* tools available. A failure to apply it properly can cause one nation to expend its resources in a well-intentioned yet flawed effort, leading to a failure to create a permanent peace. With the recent final conclusion of American involvement in Afghanistan, there is a great deal of uncertainty regarding what form economic reconstruction efforts will take in the future.

Political Transformation

---

[175] Mattox, *Saint Augustine*, 175.
[176] Michael Walzer, *Arguing about War* (New Haven, CT: Yale University Press, 2004), 76.

It is not enough that a country be rebuilt and that its leaders be held accountable. The same political structures that gave rise to tyranny in the past can do so again if not reformed. Returning to the previous example of Napoleon I, he was eventually removed by placing him in exile. However, the European powers replaced him with the same Bourbon dynasty that had been toppled by the French Revolution.[177] This decision only led to greater unrest that eventually led to the same dynasty being removed from power again. At the end of the First World War, nations experienced political transformation, but it was often an internal, rather than an external phenomenon. In Germany, Kaiser Wilhelm was forced to abdicate by his own people and the nation transitioned to the Weimar Republic and, in Russia, the czar abdicated, and, after a series of civil wars, the nation became the Soviet Union.[178] It was apparent, though, that these changes created further problems on the international stage.

At the end of the Second World War, the Allies took charge of the governments of Germany and Japan in an effort to create more democratic, West-leaning political structures. In Germany, the members of the Flensburg Government, which was the successor to Hitler's government in Berlin, were arrested.[179] The Allies kept Germany divided into four occupied zones and, years later, the western part of Germany became the Federal Republic of Germany.[180] In other words, with the exception of the Soviet occupation of East Germany, which was another matter, West Germany was occupied until it was ready to become a thriving democracy.

There were some differences in Japan. Under General MacArthur's military occupation, the Emperor of Japan retained his title but lost his authority and become a permanent figurehead.[181] The military dictatorship that had questionably ruled in the emperor's stead during the war was abolished. Japan was forced to adopt a new constitution and, as with Germany, the Allied occupation ended when the nation was ready to become

---

[177] Markham, *Road to St. Helena*, 119.
[178] Liudmila Novikova, *An Anti-Bolshevik Alternative: The White Movement and the Civil War in the Russian North*, trans. Seth Bernstein (Madison, WI: University of Wisconsin Press, 2018), 8.
[179] Detlef Junker, Philipp Gassert, Wilfried Mausbach, and David B. Morris, *The United States and Germany in the Era of the Cold War, 1945-1990* (Cambridge, UK: Cambridge University Press, 2004), 50-51.
[180] Ibid.
[181] Stephen Large, *Emperor Hirohito and Showa Japan: A Political Biography* (New York, NY: Routledge, 1992), 135.

a stable democracy. By all accounts, these efforts were quite successful as Germany and Japan have remained politically stable.

However, later attempts to enact Western-style democracies met with mixed success. In Iraq, after Saddam Hussein was removed from power, the United States and its allies did attempt to create a parliamentary system.[182] That system does remain in place to this day, but sectarian tensions in Iraq make that government's long-term viability questionable, at best. The issue is that Iraq is home to more than one people group and dominant faith. The Shiite Arabs in the south have long-standing tensions with the Sunni Arabs near Baghdad. In addition, the Kurds in the north have had problematic relations with the rest of the country and, in fact, seek to create their own state.[183] To add to the problem, several militias continue to operate outside of government authority.

Efforts in Afghanistan have been even more problematic. After the removal of the Taliban from power in 2001, America and its allies managed to install a new government in Kabul. However, given Afghanistan's tribal nature, the government had limited real authority outside of the capital.[184] With the exit of the Western powers from the country and the return of the Taliban to Kabul in 2021, there is an appearance of central authority as the Taliban was able to take control of the local governmental structures in each of the provinces first. However, it appears that they are having considerable difficulty with restoring infrastructure, maintaining the economy, and garnering international support.

This period of trial and development of *jus post bellum* political reform demonstrates that these types of reforms are possible and are necessary. However, as with other aspects of the *jus post bellum* framework, the actual application of these political reforms can have serious limitations. Furthermore, as the world moves further into the twenty-first century, several questions remain. The end goal of political reform remains the same, though. How can a nation ensure that a belligerent nation will remain at peace?

Critically, the victorious nation must decide when to depart the defeated nation and allow its new government to stand by itself. As Klein

---

[182] Ali A. Allawi, *The Occupation of Iraq: Winning the War, Losing the Peace* (New Haven, CT: Yale University Press, 2007), 283.
[183] Ibid., 73.
[184] Thomas Barfield, *Afghanistan: A Cultural and Political History* (Princeton, NJ: Princeton University Press, 2010), 53.

writes, "To achieve this goal of returning sovereignty to the vanquished, there should be a phase of *jus post bellum* to complete the transition from control and security established under the belligerent occupation to a just peace."[185] It is a given that some government, placed upon the belligerent by the victor, must exist in order to maintain stability after the conflict has ended. It is also reasonable to assume that the victor must depart at some point and return sovereignty to the newly reformed nation. The difficulty lies in deciding exactly the correct time to execute this transition.

In the cases of Germany and Japan, political reformation took several years but, when the occupying forces departed, a lasting peace ensued and the government structures that were created at the time have ensured to this very day. In Iraq, though, the new government has lasted for several years following the departure of allied soldiers, but it remains somewhat weak. In Afghanistan, of course, twenty years of occupation never brought about a stable political system.

Two issues in this *jus post bellum* effort are immediately apparent. In the nations where political reformation has failed, combat operations did not cease before the commencement of a reconstruction effort. Germany and Japan had surrendered while, in other nations, combat and nation-building occurred simultaneously. Second, Germany and Japan contained singular, unified people groups. Iraqis and Afghans have never been truly unified, relying instead on tribal structures, thus creating serious challenges for any effort to unite the people under one democratic government.

In conclusion, *jus post bellum* efforts to reform governmental structures have, at times, proven to be successful and, therefore, have seemed to support the Augustinian ideal of producing a lasting peace. However, the conditions at the end of the war may not allow a smooth transition, especially if a low-intensity conflict or a serious sectarian divide still persists. To place these issues in an Augustinian just war theory context, efforts to create an immediate peace have been at cross purposes with plans to ensure a lasting peace. Put simply, plans to embed safe, stable democracies in these nations may have unworkable and, from an Augustinian just war perspective, unnecessary. As Bass notes, "The task is to create a non-genocidal society, not a perfect one."[186] It is possible that Augustine would have approved of governmental reform, but he may have also recognized the limits of this type

---

[185] Klein, "Post Conflict Peace," 173.
[186] Bass, "Jus Post Bellum," 403.

of reform and would have found any conditions that create a lasting peace to be acceptable.

Humanitarian Aid

Provision for humanitarian aid is possibly the most overtly humane aspect of the *jus post bellum* framework. Hugo Slim defines humanitarian aid as "a compassionate response to extreme and particular forms of suffering arising from organized human violence and natural disaster."[187] Indeed, it seems morally right to provide aid and comfort through means such as the provision of food and medicine to victims of war. In any war, even when just war principles are followed, civilian displacement, starvation, and suffering are inevitable. Therefore, as a part of any conclusion of a conflict, aid must be rushed to these people.

Yet, some consideration must take place before providing this aid. It is important that the people get what they need, but it is imperative that the providing nation ensure that aid is distributed properly. As Slim writes, "Trying to help other people is a very good thing to do, but it is not always an easy thing to do. Because it is difficult, helping can go wrong. Helping people requires some kind of access to them and a certain freedom of operation."[188] Given that foreign aid is expensive, it is also important that the donating nation decide when that aid should end.

It is possible to trace the development of nation-supported humanitarian aid back to the establishment of non-governmental organizations (NGOs), such as the Red Cross, in the nineteenth century.[189] The size and number of these organizations greatly expanded in the early twentieth century in response to the First World War and the civil wars in Russia.[190] By the middle of the twentieth century, the provision of humanitarian aid had become an international norm. First, large, global assistance organizations had come into being. As Slim notes, "The creation of specific UN agencies like UNICEF, UNHCR, and the World Food Programme set in train the modern inter-state practice of humanitarian

---

[187] Hugo Slim, *Humanitarian Ethics: A Guide to the Morality of Aid in War and Disaster* (Oxford, UK: Oxford University Press, 2015), 1.
[188] Slim, *Humanitarian Ethics*, 2.
[189] David P. Forsythe and Barbara Ann J. Rieffer-Flanagan, *The International Committee of the Red Cross: A Neutral State Actor* (New York, NY: Routledge, 2007), 6.
[190] Slim, *Humanitarian Ethics*, 39.

action."[191] Second, humanitarian efforts had become encoded into international law.[192]

Also, as part of its rebuilding efforts following the Second World War, the United States sought to create a permanent, well-funded, governmental humanitarian organization. Eventually, various efforts were combined into the Agency for International Development (USAID).[193] Since that time, USAID's mission has distributed billions of dollars in humanitarian relief to nations discussed in this work, such as Iraq and Afghanistan.[194]

For sake of clarity, it is important to note that the relief efforts of the United Nations and of non-governmental humanitarian organizations do not necessarily operate outside of governmental control.[195] These organizations receive governmental funding and generally operate under the protection of a host government. It is also important to note that non-governmental organizations can work alongside fully government-sponsored aid organizations.[196] Furthermore, since the mid-twentieth century, nations have used military assets to provide humanitarian aid after a conflict as well as disaster relief.

For purposes of deciding who is to receive humanitarian assistance, especially during wartime, a set of legal categories has been created. As Slim notes, "In the development of international law, these humane principles have been shaped as particular duties, rights, and rules or constructed into specific moral identities like non-combatant, civilian and refugee. These moral and legal norms have been derived as specific secondary principles to regulate the humane conduct of armed conflict."[197] Non-combatants, civilians, and refugees are considered to be outside the boundaries of acceptable violence during wartime, though some military personnel, such as chaplains and medics, are to be left unmolested during a conflict.[198] Refugees

---

[191] Ibid.
[192] Ibid.
[193] Rachel M. McLeary, *Global Compassion: Private Voluntary Organizations and U.S. Foreign Policy Since 1939* (New York, NY: Oxford University Press, 2009), 89-90.
[194] McLeary, *Global Compassion*, 32-33.
[195] Shai Dromi, *Above the Fray: The Red Cross and the Making of the Humanitarian NGO Sector* (Chicago, IL: University of Chicago Press, 2020), 2.
[196] Ibid., 138.
[197] Slim, *Humanitarian Ethics*, 52.
[198] Michael Bothe, K.J. Partsch, and W.A. Solf, *New Rules for Victims of Armed Conflicts: Commentary on the Two 1977 Protocols Additional to the Geneva Conventions of 1949* (Boston, MA: Brill, 2013), 94.

are essentially the same as civilians, but the fact that they are fleeing means that they will have extraordinary needs.[199]

Humanitarian aid is also distinct because, while it may appear to be similar to other forms of aid, the circumstances differentiate it from other types of aid that could be provided during wartime. Medical aid to soldiers would not be considered humanitarian aid but similar aid to refugees would be "morally distinct from other kinds of aid in armed conflict."[200] The various types of humanitarian aid have become quite diverse. From a material standpoint, this aid typically involves food, water, healthcare, and the essentials required to maintain a decent quality of life.[201] At a very basic level, all people require access to food and healthcare. The idea is that the prevention of suffering encompasses more than simply meeting physical needs but also ensuring that the person has a decent quality of life.

Providing material comfort, however, is not enough. Just as people have needs for food and medical care, they also need a safe environment. Humanitarian protection "recognizes types of aid that help to keep people safe from violence and degrading treatment and connected with their families, like communication technology, protective buildings, identity cards and prison visiting."[202] Through advocacy, humanitarian support has been tied to the *jus post bellum* principle of justice. Humanitarian advocacy "impartially and neutrally draws attention to the needs of vulnerable populations and violations of international law."[203] As noted earlier, it is important that efforts be made to provide restitution for those who have been harmed by war.

In summary, the provision of humanitarian aid is a *jus post bellum* line of effort that has seen tremendous growth in the last century. This type of effort is not limited to a post-war environment and, in fact, often takes place after a natural disaster. However, for the purposes of this study, post-war humanitarian aid is intended to alleviate suffering and restore a sense of peace to an affected post-war area. Unlike other *just post bellum* principles, it comes from a variety of sources, each with its own budget and agenda, thus creating a unique set of challenges. However, the overriding Augustinian principle remains. Humanitarian aid is a basic, necessary function of Christian leadership that is intended to restore lasting peace to the people whom most

---

[199] Iasiello, "*Jus Post Bellum*," 45.
[200] Slim, *Humanitarian Ethics*, 55.
[201] Pattison, "*Jus Post Bellum*," 649.
[202] Slim, *Humanitarian Ethics*, 55.
[203] Ibid.

need it. As Iasiello writes, "While Augustine is usually associated with the formulation of the just war theory as we know it, he was also a pastor, and as such he was concerned with war's impact on people as he was with defining the parameters of a just war."[204] Even with the secularization of just war theory, this principle of a moral duty to render aid remains valid.

---

[204] Iasiello, *"Jus Post Bellum,"* 49.

4. *JUS POST BELLUM* IN PRACTICE

This study has provided a high-level overview of Augustinian thought as related to war and has traced the development of Augustine's just war principles from his own day until the present. This work has also discussed the formulation, refinement, and challenges of the relatively new *jus post bellum* framework while asserting that this framework has served as the means by which nations have attempted to fulfill the just war goal of a lasting peace.

Though just war theory had moved from being a sacred to a secular construct, its critical components remained and could be traced directly back to Augustine. When the ideals of *jus post bellum* were developed, they, too, were attached to Augustinian ideals even though they had not seen the same level of long-term development as other pillars of just war theory. Still, the central question posed in this study remains. Given that *jus post bellum* has been retroactively attributed to Augustine, would he have approved of it or, to put this another way, have *jus post bellum* theorists assumed that their writings could be correlated with Augustinian thought?

This chapter contains the data that will answer this important question. It will examine, in detail, many of the post-war *jus post bellum* activities that have taken place in the recent past. For reasons of scope, the areas discussed will be American involvement in Germany and Japan after the Second World War, in Iraq after the Second Gulf War, and in Afghanistan during the Global War on Terror. This data will be presented systematically with each period's *jus post bellum* efforts being divided into the general *jus post bellum* lines of effort discussed previously in this work. The end result will be a nuanced conclusion that will find that Augustine would have approved of many, but not all, of these recent post-war efforts.

Post-World War II Reconstruction of Japan and Germany

By the end of the Second World War, the social, economic, and political situation in Europe and in Asia was dire. Germany and Japan, especially, had been devastated by years of strategic bombing campaigns that had systematically targeted ports, cities, and critical infrastructure. Now that the war had ended, the question was how to avoid making the same post-First World War mistakes that had laid the groundwork for a renewed conflict within a generation. The end result was a concentrated effort to hold war criminals accountable, rebuild economies, conduct political reform, and provide necessary humanitarian aid.

War Crime Trials and Accountability

The International Military Tribunal, held in Nuremberg from 1945 to 1946, convened in order to adjudicate the indictments of German leaders for war crimes and for crimes against humanity. The list of crimes included actions such as religious persecution, murder, enslavement, and extermination. Notable defendants were the heads of each of the German armed forces and the senior Nazi party officials who were considered to be within Hitler's inner circle. Key pieces of evidence at the trial included photographic evidence from the concentration camps that were liberated towards the end of the war and Nazi propaganda films in which the party had discussed its own plans to enact further atrocities.[205] Most of Hitler's key staff, who were the first to be tried, were found guilty. Approximately half faced execution while the remainder of the convicted men were given lengthy prison sentences.

In an analysis of the efficacy of the trial, several key components are relevant. First, did the tribunal actually try all of the suspected Nazi war criminals? Unfortunately, though several of the most notorious of Hitler's followers were tried, the vast majority of those who committed crimes went unpunished. Kevin Heller notes that "of the 2,500 'major war criminals' identified...only 177 ever stood trial – a mere 7 percent."[206] Part of the issue

---

[205] Bob Carruthers, ed., *The Gestapo on Trial: Evidence from Nuremberg* (South Yorkshire, UK: Pen and Sword Books, 2014), 66-67.
[206] Kevin Jon Heller, *The Nuremberg Military Tribunals and the Origins of International Criminal Law* (New York, NY: Oxford University Press, 2011), 370.

was that some potential defendants were unavailable because they died at the end of the war or fled to friendly nations. Others were not prosecuted due to political reasons or reasons of national interest. As Heller notes, "That retributive shortfall would have been troubling even if 177 defendants had represented the most major of the 'major war criminals.' But that seems unlikely given how many important suspects were not prosecuted solely for financial, temporal, or logistical reasons."[207] To put this another way, the vast majority of suspected war criminals went unpunished.

Second, was the tribunal simply another post-war show trial, or was it a fair, realistic legal mechanism? The overwhelming evidence is that those who were convicted truly were guilty men. Heller writes, "The convictions themselves appear to have been retributively just. There is no obvious example of a defendant in the trials who should have been acquitted but was not, and the tribunals acquitted 35 of the 177 defendants."[208] Considering the length and complexity of the trial, the amount of evidence, and the potential biases against the accused, the fact that so many were acquitted is quite remarkable.

This leads to another question. Is it possible that guilty men were acquitted in spite of their guilt? In fact, it is possible that this phenomenon occurred. Heller notes, "There is no question that the reliability of the OCC's selection process was undermined by the office's lack of time and personnel, the amount of evidence that the attorneys and analysts had to process, and the questionable reliability of some of that evidence."[209] He makes an interesting point. Though the tribunal was large in size and scope, it took place relatively soon after the war's conclusion. In Germany, the war ended in May 1945 and the tribunal began that same year. Processing the evidence and preparing for trial would have been exceedingly difficult.

Finally, was the tribunal actually worth the effort? It is evident that relatively few criminals were actually brought to justice. Though the tribunal appeared to have been conducted fairly, it may have commenced too quickly to give prosecutors adequate time to prepare their cases. In Heller's assessment, despite these issues, the tribunal was worthwhile for "at least attempting to impose fair and consistent sentences." [210] To understand Heller's optimism, it is important to place the Nuremberg Military Tribunal within its historical context. Until that point in history, very few major wars

---

[207] Ibid.
[208] Ibid.
[209] Heller, *The Nuremberg Military Tribunals*, 370.
[210] Ibid. 371.

had ended with fair, consistent trials for the belligerents. The tribunal, while imperfect, was possibly the best *jus post bellum* effort to provide justice until that time.

The Tokyo War Crimes Tribunal began shortly after the war in the Pacific ended and, in some ways, mirrored what was taking place in Nuremberg. It was the hope of those involved that this tribunal would have the same impact on international law. As Robert Cryer and Neil Boister write, "The entrenching of the Nuremberg precedent and contributing to international law was on the minds of the Tokyo Charter's drafters. It was also on the minds of a number of judges."[211] However, for a variety of reasons, the Tokyo International Military Tribunal has not received a high level of attention outside of Japan since its conclusion.

Part of the issue was that the tribunal assumed that Japan's desire to wage war in China and in the Pacific was simply out of a sense of aggression when, for the Japanese people, the story was much more nuanced. As Cryer and Boister note, "The indictment and trial reflected a particular view of Japan's recent history held by some among the Allies."[212] Many factors were considered as Japan formulated its wartime aims, to include earning greater access to raw materials and removing colonial rule from Asia.[213] Second, the rules of evidence used during the tribunal had the effect of preventing some crimes from being brought to the forefront of the conversation. The tribunal heavily favored "the admission of documentary evidence over live testimony."[214] While physical evidence can be quite valuable, it allowed the tribunal to largely overlook those crimes in which documentary evidence is lacking.

Third, except for those few leaders who were executed as a result of the tribunal, most of those who were responsible or potentially responsible for war crimes received little to no punishment. The emperor, for example, was never charged with a crime, perhaps establishing his permanent role as a figurehead.[215] Of those who were sent to prison, all were released within ten years, and many went on to successful careers in governmental service

---

[211] Robert Cryer and Neil Boister, *The Tokyo International Military Tribunal – A Reappraisal* (Oxford, UK: Oxford University Press, 2008, 302.
[212] Ibid., 312.
[213] Aaron Stephen Moore, *Constructing East Asia: Technology, Ideology, and Empire in Japan's Wartime Era, 1931-1945* (Stanford, CA: Stanford University Press, 2013), 52.
[214] Cryer and Boister, *Tokyo International Military Tribunal*, 314.
[215] Madoka Futamara, "Individual and Collective Guilt: Post-War Japan and the Tokyo War Crimes Tribunal," *European Review* 14, no. 4 (2006): 474.

afterward.[216] An additional contributing factor was an unwillingness on the part of the Chinese government to become involved in these war crime trials. Given that most suspected crimes took place in China, this was a political decision on the part of China's new communist government. The issue was that any attempt to hold Japanese officials accountable would have required collaboration between two opposing sides in China's civil war.[217]

With these points in mind, would it be fair to say that the Tokyo Military Tribunal, as a *jus post bellum* effort was successful? It is unquestionable that it was a flawed process, perhaps even more so than the Nuremberg trials. From an Augustinian just war perspective, though, it may be accurate to say that these first attempts at pre-war justice were a step in the right direction towards providing justice in furtherance of a permanent peace. The trials themselves represented efforts to hold criminals accountable through the use of fair processes in which rights were respected. The punishments levied were an indication that justice could be applied fairly. Furthermore, given that names such as "Nuremberg" are watchwords for postwar criminal trials, it may be fair to argue that the desired outcome of deterrence has been achieved.

Economic Reconstruction

The postwar economic reconstruction of Japan and Germany took two different forms. In Japan, General MacArthur became a virtual dictator who used the resources at his disposal in the Pacific to help rebuild the nation as quickly as possible.[218] In Europe, the Marshall Plan, formally known as the European Recovery Plan, provided $13 billion in economic aid for the purposes of rebuilding transportation, agriculture, and factories as well as cities.[219] Notably, Germany and Italy received a high level of attention as well as those nations, such as Greece, that were directly threatened by rising Soviet influence.

The Marshall Plan was unquestionably a *jus post bellum* effort with its roots in the aftermath of the First World War. As Michael Holm writes, "It began not in the spring of 1947 as most scholars insist. Its short-term origins

---

[216] Cryer and Boister, *Tokyo International Military Tribunal*, 316-317.
[217] Ibid., 324.
[218] Richard B. Finn, *Winners in Peace: MacArthur, Yoshida, and Postwar Japan* (Los Angeles: University of California Press, 1992), 44.
[219] Michael J. Hogan, *The Marshall Plan: America, Britain, and the Reconstruction of Western Europe, 1947-1952* (Cambridge, UK: Cambridge University Press, 1987), 30.

went back at least to the spring of 1919."[220] When Woodrow Wilson participated in the Paris Peace Conference, he was unique among the allied leaders because he sought to create a permanent international order that would prevent such a war from occurring again. Notably, he also refrained from any desire to be overly punitive towards Germany and Austria-Hungary. Unfortunately for him, his main idea, the League of Nations, failed because the treaty was not ratified back in America. Yet, the principal goal of maintaining permanent peace in Europe remained.

In some ways, the ideals behind the Marshall Plan had their genesis deep within American culture. Holm notes, "The Marshall Plan was never a singular endeavor created solely for the purpose of European Reconstruction. It had its origins in America's recent past and a powerfully shared belief that the United States possessed a responsibility to uplift a world in desperate need of guidance"[221] Altruistically, America felt a duty to help the world during a time when it was, perhaps, the only nation that had the capability to do so. Furthermore, the success of the economic relief effort gave Americans the sense that the people and their government could work together as a force for good. As Holm writes, "In the decades that followed, the ideas, desires, and accomplishments of the Marshall Plan assumed almost a legendary status. In the United States, it allowed Americans to remember the government as a source of good."[222] In other words, as a *jus post bellum* effort, the Marshall Plan has been considered to be not only a tremendous success but also a representative example of national pride.

Though America has deserved a great deal of credit for its ability to actualize and execute this plan, much of the work was done by the European nations themselves. They utilized the aid provided and rebuilt infrastructure, industry, and cities. By the conclusion of the effort, "the overall industrial production aggregate among recipient nations was 41 percent above 1938 levels, while the GNP rose some 25 percent."[223] What was, perhaps, more remarkable was that this period of economic reconstruction unified Western Europe to a degree never seen before. As Holm notes, "What is worthy of reflection is the fact that the Western Europe that emerged after the Second World War was more united than ever before, to such an extent that today's students take the concept of *Europe* for granted."[224] As part of this outcome,

---

[220] Michael Holm, *The Marshall Plan: A New Deal for Europe* (New York, NY: Routledge, 2017), 1.
[221] Ibid.
[222] Ibid., 117.
[223] Ibid., 115.
[224] Ibid., 129.

though the nations of Western Europe have been involved in wars since that time, they have been at peace with each other for almost eighty years.

The *jus post bellum* economic reconstruction effort in Japan was, perhaps, more daunting. Japan had been subject to a continual strategic bombing campaign that was intended to make an invasion of its home islands unnecessary. William Leavitt writes, "Every major city was described as a 'wilderness of rubble.' Between two million and three million people had lost their lives."[225] Furthermore, the nation's economy was ruined during a time when winter was just about to set in. Leavitt writes that "there was little phone or train service, and virtually no power plants were in operation. Cities and factories were gutted, and the entire population faced starvation."[226] Fortunately for Japan, America had given MacArthur the authority and the resources that he needed to provide relief extremely quickly. Of note, MacArthur had full control over the Japanese government, and, in addition, retained the goodwill of the American government and access to the stockpiles of American equipment remaining in the Pacific.

MacArthur's efforts were wide-reaching and encompassed every area of Japanese society. In terms of economic reconstruction, he delivered economic aid as was done in Europe. In addition, he made two major attempts at creating lasting reform. First, he tried to reform an agricultural system that was still locked into a feudal-type series of arrangements. Before the war, most Japanese farmers lived on a subsistence level and paid rent to large landowners. Leavitt writes, "The plight of peasants in Japan was a wretched one. MacArthur demanded far-reaching change. Absentee landlords were expropriated, and their lands purchased by the government and sold to tenants at prewar prices."[227] Land reform was about more than simply providing justice for the peasant class. The idea was to invigorate the middle class so that these people would have a greater degree of political power and, thus, become more influential contributors within Japanese society.

Second, he attempted to break the dominance of the *zaibatsu* which, before the war, consisted of only a few companies yet managed to dominate Japanese industry.[228] Briefly, *zaibatsu* were vertically integrated businesses that

---

[225] William M. Leavitt, "General Douglas MacArthur: Supreme Public Administrator of Post-World War II Japan," *Public Administration Review* 75, no. 2 (March/April 2015): 315.
[226] Ibid., 315-316.
[227] Ibid., 322.
[228] Finn, *Winners in Peace*, 199-200.

tended to have monopolistic power over a sector of the economy akin to what Rockefeller accomplished with Standard Oil in an American context. MacArthur, in an effort to allow Japanese citizens greater access to the economy, was initially successful in breaking up the *zaibatsu* system. However, large conglomerates reformed shortly afterward and have continued to drive the Japanese economy to the present day, albeit under somewhat different circumstances. As Leavitt notes, "The Japanese economy was still dominated by a handful of massive, furiously competing, yet also cooperating mega-economic giants."[229] Still, given the current positive state of Japan's economy, it is reasonable to argue that these conglomerates have had a positive impact.

In summary, post-war economic reconstruction efforts were an overwhelming success. Both Europe and Japan rebuilt their economies and quickly exceeded their pre-war industrial and agricultural output. More importantly, the nations involved have become safe, stable democracies. The financial cost to America was considerable at the time, but its *jus post bellum* efforts seem to have borne fruit. To place these efforts in an Augustinian context, it is likely that Augustine would have approved of them despite the cost providing that the American government could continue to ensure stability at home. If any criticism can be laid upon these efforts, it is that they may have been too successful. As later history will prove, applying these same templates in other nations could produce very different results.

Political Transformation

In Germany and in Japan, the need for political reform was obvious. In both cases, the previous governmental structures had failed, causing these nations to move towards belligerent status. The challenge was to introduce or re-introduce democratic governmental structures in such a way that they could maintain popular support in the years following the Allied occupation. For purposes of this discussion, changes in the German government will be held to what occurred in West Germany as East Germany remained under strong Soviet influence until the end of the Cold War.

In West Germany, the Allied occupation lasted until 1955. In the intervening years, military control slowly transitioned to civilian control before the nation was allowed to govern itself as a democracy.[230] Full independence, though, did not come until German reunification at the end of the Cold War. During this period, any association with the Nazi party was

---

[229] Leavitt, "General Douglas MacArthur," 322.
[230] Holm, *The Marshall Plan*, 28.

strictly outlawed and German rearmament was kept to a minimum. This created other problems, though, as West Germany was essentially on the front line with an expansive Soviet Union. Therefore, the North Atlantic Treaty Organization (NATO) was created in order to keep a united Europe prepared for a potential conflict with the Soviet Union.[231] Furthermore, a significant number of American forces were required to be stationed in Germany and, in fact, remain to the present day. German political reform, therefore, as a *jus post bellum* effort could be considered successful, though it was a success that took several decades to become fully realized.

Perhaps the most important questions concerned German reunification and rearmament. In a *jus post bellum* environment, what does a victorious nation do when a defeated state serves as a buffer between a much larger threat? NATO was a complex answer to this issue. As John Ikenberry writes, "The solution to German rearmament and statehood was its further integration in European economic institutions and the Atlantic alliance. A powerful and independent Germany, able to balance between East and West, was unacceptable to the United States and to other western governments."[232] Since Germany could not be allowed to rearm, essentially meaning that it could not defend itself, the requirement was for the other surrounding nations to work together for common defense. Perhaps the most noteworthy portion of the treaty is Article Five, which binds the signatories to common defense by all if any one signatory is attacked.

NATO had an additional benefit that aided European unity. Ikenberry notes, "The NATO agreement was a continuation of the Marshall Plan strategy: to extend assistance to Europe in order to improve the chances that Europe would succeed in reviving and integrating itself."[233] The allied nations came together during the implementation of the Marshall Plan. Since they were already working together and since the treaty required a strong sense of European unity in order to function, the process of working together militarily only brought the nations of Western Europe closer.

The situation was much different in Japan. As with Germany, a key concern was ensuring that the political structures within the Japanese government never allowed the nation to become a threat to peace in the

---

[231] Ibid., 108.
[232] G. John Ikenberry, *After Victory: Institutions, Strategic Restraint, and the Rebuilding of Order after Major Wars* (Princeton, NJ: Princeton University Press, 2001), 197.
[233] Ibid.

region.[234] However, since Japan had never been under a Western-style democratic government, putting such a government in place, and keeping it relevant after the American occupation, seemed to be a daunting challenge. When MacArthur was given the authority to be a military governor over the nation, he sought to impose democratic reforms rapidly yet carefully. This was aided by the fact that he was given the authority to do as he wished. Leavitt writes, "MacArthur understood that he needed to act quickly to create stability in Japan as his first priority, which is why he determined that the emperor system must stay and refused to allow interference from the Allied Council."[235] The Japanese deeply revered the emperor but, since it was likely that he had not commanded true authority in some time, it made sense to keep him in place yet powerless.

Politically, perhaps the most powerful reform was the new Japanese constitution.[236] MacArthur insisted that it include several key provisions, including women's suffrage, labor protections, and educational reforms.[237] In a sense, the constitution was to contain many of the liberal reforms that America had recently granted itself. The idea was that the Japanese people would be better educated, have greater control over their lives, and have an enhanced ability to participate in their democracy.

Several other reforms, as provided by the new constitution, are noteworthy. As Leavitt writes, "Feudal institutions, such as the peerage, had to be abolished and a representative legislature established. The budget was to be patterned after the British system. And, finally, war and the possession of armed forces were to be proscribed."[238] The constitution essentially created a parliamentary democracy in which all were equal under the law. The emperor still retained his role as the ceremonial head of state, but the people were able to select those who governed them.

From a *jus post bellum* standpoint, one of the most important reforms under the Japanese constitution was a prohibition against maintaining a large standing military force. Japan has been able to utilize a self-defense force since that time, but it has been unable to re-establish the same size military that it supported during the war. This rule was put in place as another means to ensure that the Japanese military services would never become significant

---

[234] Michael Schaller, *The American Occupation of Japan: The Origin of the Cold War in Asia* (New York, NY: Oxford University Press, 1987), 41-42.
[235] Leavitt, "General Douglas MacArthur," 320.
[236] Ibid., 321
[237] Ibid.
[238] Ibid., 321.

political stakeholders again. The end result was that, as a *jus post bellum* effort, the political reconstruction of Japan was a resounding success. The reforms imposed upon the Japanese people were new to that nation, though they had been tried elsewhere. Still, liberal reforms such as labor laws and universal suffrage were widely accepted and remain part of the Japanese political system to the present day. Furthermore, the Japanese constitution, as established during the American occupation remains in effect.

In summary, *jus post bellum* attempts to reform the political systems of Germany and Japan after the Second World War were complex and challenging, yet very successful. Reforms in Japan moved very quickly, though it is important to remember that the situation in which they were imposed was quite unique. In Europe, political reforms moved slowly, yet effectively. From an Augustinian just war standpoint, this political calculus seems acceptable provided that the long-term goals were achieved. Given Augustine's historical context within the latter days of the Western Roman Empire, he would have understood that alliances and political compromises were necessary in order to promote external order. NATO, as an organization, was established to provide that necessary balance in twentieth-century Europe. Still, NATO's mutual defense provisions have some similarities to those present at the beginning of the First World War. In other words, *jus post bellum* political reforms have successfully kept Western Europe at peace, but whether or not longer-term peace is a possibility remains to be seen.

Humanitarian Aid

Planning for relief efforts in Europe coincided with the creation of the United Nations. Even before hostilities had fully ended, humanitarian aid was being delivered into the war-torn areas of Western Europe and, eventually, efforts were expanded into parts of Asia. The major relief effort was provided by the United Nations Relief and Rehabilitation Administration (UNRRA), and the planning of this effort stretched back several years to the earliest conferences held by the Allies.[239] Therefore, since the UNRRA was a direct result of the post-war planning process, its success or failure as a *jus post bellum* line of effort is relevant to this work.

As the Axis powers were in retreat towards the end of the war, they left behind devastated areas in other nations as well as in their own. Furthermore, Germany contained a large number of prisoners who, after

---

[239] David Mayers, *America and the Postwar World: Remaking International Society, 1945-1956* (London, UK: Routledge, 2018), 13.

liberation, needed short-term care and to be returned to their homeland. The UNRRA, according to Susan Armstrong-Reid and David Murray, "Represented a promise to victims of war that, once the Axis yoke was broken, medicines and clothing and other emergency supplies would be quickly sent to rebuild shattered lives and war-torn economies."[240] The provision of food, shelter, and medicine, of course, was necessary. The greater challenge, perhaps, involved taking proper care of displaced populations as this involved close coordination with host nations.

Overall, the UNRRA was a successful effort. As Armstrong-Reid and Murray state, "Eventually UNRRA would administer $4 billion in aid. By early 1946, it was delivering essential relief supplies to over twenty countries on a scale that surpassed even the movement of munitions by Allied forces during the war."[241] This was an impressive feat, though it is also important to note that this type of aid was quite different from the economic efforts of the Marshall Plan. While the Marshall Plan was a dedicated, multi-year development plan, the UNRRA was more focused on meeting immediate needs.[242] This has much to do with the political situation. Relief was planned and executed before the war had concluded while the Marshall Plan came later in response to the threat of Soviet expansion.

While the UNRRA was a successful *jus post bellum* attempt to provide humanitarian aid, some problems with its administration became apparent. These issues will arise again and again during similar efforts later in this study. First, providing comfort to the suffering regardless of religion or ethnicity is a laudable idea, but equality in execution can be difficult. As Armstrong-Reid and Murray note, "UNRRA was handicapped in its ability to ensure that relief was distributed effectively and equitably without discrimination by race, religion, or politics as required in its mandate."[243] The UNRRA was moving food, medicine, and shelter to over a dozen nations in an extremely rapid fashion. It may have been unreasonable to assume that local politics would not have played a role in deciding who had better access to this aid.

Second, while the UNRRA was responsible for delivering aid, it depended on other nations to provide the aid and to allow its distribution. Armstrong Reid and Murray note, "Its dependence, both on previously existing Allied bodies that controlled the allocation of supplies and on the

---

[240] Susan Armstrong-Reid and David Murray, *Armies of Peace: Canada and the UNRRA Years* (Toronto, Canada: University of Toronto Press, 2008), 4.
[241] Armstrong-Reid and Murray, *Armies of Peace*, 4.
[242] Mayers, *America*, 16.
[243] Armstrong-Reid and Murray, *Armies of Peace*, 6.

recipient countries that developed relief-distribution priorities, left UNRRA in the unenviable position of being a referee without authority to impose penalties."[244] The Allied powers gave the UNRRA the authority to operate, but the actual delivery of goods depended greatly upon the local laws of each recipient nation. To put this another way, those responsible for this relief effort had very little control over their supply chain. As a result, shortages in some areas and accusations of unfairness or incompetency were inevitable.

In terms of alleviating suffering among civilians in Japan, MacArthur prepared for the pending winter by moving existing wartime supplies into the county. Leavitt writes, "Within days of beginning the occupation, MacArthur had army field kitchens set up to feed hundreds of thousands of starving people. He also seized some 3.5 million tons of food stockpiled by the American army in the Pacific in order to see Japan through the winter."[245] The end result was that starvation was prevented, thus allowing the process of economic reconstruction to continue.

In summary, the *jus post bellum* humanitarian relief efforts at the end of World War Two were successful, but this success may be difficult to replicate. In the Pacific, disarming soldiers and bringing them home was facilitated by an American shipbuilding industry that had expanded tremendously throughout the war. Efforts in Europe were similarly successful, though they came about through different means. Still, it can be argued the humanitarian relief, as planned by the Allies, allowed the longer-term reconstruction to commence and, thus, was in alignment with Augustinian just war principles.

Impact on Jus Post Bellum Development

Undoubtedly, the Allied powers sought to ensure that the mistakes made at Versailles several decades earlier were not repeated. Were these efforts successful and did they truly serve to preserve a lasting peace? This review will re-examine each of the four *jus post bellum* pillars discussed in this study, as related to post-Second World War development, with an eye toward long-term implications.

Beginning with the Nuremberg Tribunal, the evidence indicates that this aspect of post-war justice produced mixed results. As the first effort of its kind, it was, comparatively, much better than any system that denied

---

[244] Ibid., 6.
[245] Leavitt, "General Douglas MacArthur," 321.

accountability. Those who were tried for the worst of war crimes were found guilty in a fair trial and were fairly sentenced. The fact that a victorious coalition can treat its defeated enemies in such a manner appears to be remarkable.

Another benefit of the Nuremberg Tribunal is that, by its nature, it caused the Allied powers to compile and enter into an official record proof of German wartime atrocities. Heller writes, "More than 1,300 witnesses testified during the trials, the parties introduced more than 30,000 documents into evidence, and the judgments run more than 3,800 pages. In the aftermath of the trials, only the most committed apologist could maintain that the Holocaust – and the Nazis' other crimes beyond number – were 'fable, not fact'."[246] This is a very important development because one of the goals of *jus post bellum* justice is to provide some type of restitution to the victims. While it is true that the West German government could never restore what had been taken, there is value in proving that these crimes occurred to the world at large.

In Japan, the issue was that the tribunal appeared to lack legitimacy. As Cryer and Boister note, "The main lesson...must be related to the victor's justice critique. If a trial is perceived as being unfair, its intended legacy is likely to be undermined."[247] These trials focused more on criminal prosecution for starting wars rather than for crimes during war. Since the commencement of war had not been prosecuted as an individual war crime before this time, there was a sense that these trials were more about providing justice for the war's victors. Furthermore, even compared to Nuremberg, very few people received either capital punishment or lifelong prison sentences.

The economic and political reconstruction of Europe and Japan was highly successful and, perhaps, these efforts clearly demonstrated the value of these pillars of the *jus post bellum* framework. According to Leavitt, "Japan in the twenty-first century is a prosperous, democratic country with the world's third-largest economy. The path to Japan's success as a democratic nation and economic powerhouse began with the occupation of Japan in 1945."[248] A direct line can be traced from the reconstruction of a ruined Japan in 1945 to the peaceful state that it is in the present day. In Europe, the Marshall Plan became a moment of American pride. As Holm writes, "The Marshall Plan was a bright shining moment that helped fortify Americans'

---

[246] Heller, *The Nuremberg Military Tribunals*, 372.
[247] Cryer and Boister, *Tokyo International Military Tribunal*, 326.
[248] Leavitt, "General Douglas MacArthur," 315.

faith in the nation and in what the true purpose of the Cold War really was."[249] In both cases, the democratic institutions created and the alliances that were forged remain in effect.

Similar lessons will be apparent as a result of the process of political reconstruction. First, it is remarkable that Japan has maintained a constitution that was imposed upon it by a foreign power. This is a testimony to the success of the liberal reforms contained in the document and the willingness of the people to follow its precepts. In Germany, political reform was complicated but successful. The realities of the Cold War coupled with concern about German resurgence meant that Germany remained divided, but West Germany eventually transitioned to a democratic government that was supported by its own people as well as the nations of Western Europe.

Finally, the humanitarian efforts provided at the end of the war were truly remarkable and are a testimony to the necessity of helping innocent victims as a component of any *jus post bellum* effort. Direct aid had its challenges and will always have its challenges as it depends on the goodwill of all stakeholders. While humanitarian aid deployment in Europe was inefficient, the people who needed support received it. In Japan, it became apparent that a standing military can be very successful at utilizing its resources to deliver assistance rapidly at the war's conclusion. This *jus post bellum* pillar will only expand in later years to include a patchwork of public and private humanitarian endeavors.

Comparison with Augustinian Thought

If Augustine were able to observe these post-war activities, would he have approved of them? The answer is complex because the mid-twentieth century would have been far removed from his cultural context. However, it may be possible to crosswalk some of the issues that he observed to the American context at the end of the war.

In 1942, America had the manpower and the industrial capacity to enter the war for the correct reasons and produce an outcome that ended aggression in Europe and in Asia. When the war had concluded, America retained the ability to at least attempt to ensure that peace could be maintained. Augustine probably would have agreed with American entry into the war. As Patterson writes, "Just war theory's purpose was to call for responsible action while imposing limits, recognizing the moral obligation of

---

[249] Holm, *The Marshall Plan*, 117.

leaders to defend and promote order, security, and justice in a fallen world."[250] Just as the sovereign in Augustine's time was deemed responsible for justice and security both within and outside his borders, it can be argued that America had a similar responsibility. Augustine's ideal of leaving the people free to practice their Christian faith may not have been as relevant in the twentieth century but providing people with a decent level of comfort and security so that they could live their lives freely would have been a reasonable goal both in recent times and in Augustine's day.

What about conduct after the war has concluded? Patterson feels that under Augustinian thought, certain responsibilities remain after the conclusion of hostilities. He writes, "The political ethic of responsibility inherent in the just war tradition expected, and continues to expect, that sovereigns (governments) rule their dominions with justice after fighting has stopped."[251] He makes a salient point. If a nation enters a war in order to produce a just outcome, then that nation would be wrong to leave matters in disarray at the war's conclusion. Mattox makes essentially the same argument when he finds that some nation, even if imperfect, must take the initiative to restore order.[252] Furthermore, though there was no specific *jus post bellum* requirement at the time, the need to act in a moral manner should have been obvious. As Patterson notes, "The just war tradition historically did not need *a jus post bellum* because there were robust religious, moral, and philosophical teachings – from the Old Testament to Aristotle and Aquinas – about the ethics of righteous governance."[253] In a Second World War context, though, these requirements for conduct after the war would have been very complicated.

First, America was the only nation that could have possibly rebuilt not just one, but two major powers. Such an effort had never been attempted before. On a related note, the requirement for humanitarian aid was equally staggering. Third, the act of introducing new governmental structures into societies in a way that would produce lasting results carried a great deal of risk, especially in the face of an emergent nearby rival. Even the personal accountability portion of these *jus post bellum* activities had been untried, with the closest approximation being Napoleon's exile to St. Helena. In short, though it seems simple to assume that America should have performed these activities, they all carried a high level of political and financial risk. It is

---

[250] Patterson, *Ending Wars Well*, 33.
[251] Ibid.
[252] Mattox, *Augustine*, 175.
[253] Patterson, *Ending Wars Well*, 33.

possible that, perhaps, the lessons of Versailles convinced the American people that the risks were justified.

That being said, under Augustinian thought, America was morally correct to at least attempt to perform each of these activities because it had both the capacity and the duty to restore peace.[254] In terms of enforcing standards of justice, both individually and collectively, establishing tribunals and attempting to hold legally impartial trials was the correct approach. As Patterson states regarding Augustine's stance on the subject, the establishment of order is a "moral project to reflect the values of justice and ethics found in the City of God."[255] Order and justice are intertwined. Even if the earthly city can never truly be the equal of the City of God, a righteous government would necessarily attempt to make the closest possible approximation.[256] The fact that this type of postwar justice was relatively new did not mean that it was morally wrong. It may be better to say that fair postwar trials should have been conducted long before this point in history.

In terms of spending a vast fortune with the goal of rebuilding economies and providing humanitarian aid, America was morally correct in making these attempts. Under Augustinian thought, financial expense, to an extent, is divorced from morality. As Patterson states, "Augustine argued that society was rooted in a moral order, not simply the economic interactions of human beings."[257] Of course, if a nation were to bankrupt its own people in the hope of rebuilding another, that would be a different matter. The sovereign or government retains responsibility for internal welfare. Though America spent heavily at the end of the Second World War, it was not in danger of bankruptcy. The effect was just the opposite as these rebuilding and humanitarian efforts re-established free trade between nations.

Finally, attempting to create liberal, democratic institutions was also a worthwhile task. After all, it is reasonable to assume that Augustine would have approved of replacing tyrants with systems of government that respected the right of the people. Changing the political order frees the people of a nation to pursue goals that promote peace and tranquility within their own borders.[258] Taking steps to ensure that the people accepted and embraced these new governmental structures was also morally correct.

---

[254] Mattox, *Augustine*, 177.
[255] Patterson, *Ending Wars Well*, 43.
[256] Ibid., 42.
[257] Ibid.
[258] Ibid., 43.

America's *jus post bellum* conduct after the Second World War was imperfect. First, political considerations weighed heavily in the calculus for justice, economic reconstruction, and political reforms. Not all war criminals were tried and, in fact, many avoided trials because they could not be found or were useful in the pending arms race with the Soviet Union. A whole treaty organization was built around the concern that Germany must be kept weak while, at the same time, serving as a bulwark against Soviet expansion. Furthermore, the Marshall Plan was as much about halting blocking Soviet influence as it was about rebuilding nations. Still, it is reasonable to assume that even Augustine would not have been surprised by these kinds of realities. As discussed earlier in this study, he was not blind to the politics of his day. It is enough that America and its allies successfully rebuilt the war-torn nations of the Second World War in such a way that many of the affected regions quickly became peaceful and prosperous.

Second, the *jus post bellum* efforts themselves were, in many cases, either inefficient or difficult to replicate. While an untold amount of humanitarian aid was delivered to Europe, it was not uniformly delivered. Furthermore, the UNRRA found itself in a challenging situation as it had to answer to a wide number of governing bodies. The war crime trials held many accountable and seemed to be conducted fairly, but regulations regarding rehabilitation and parole ensured that prison terms tended to be shorter than expected. The economic reconstruction of Japan was extremely successful, but it relied upon the presence of a virtual dictator who happened to be both competent and benevolent.

However, given the fact that America accepted such a high degree of risk and attempted ideas that were novel at the time, it would be fair to say that these inefficiencies could be forgiven. Most importantly, under the Augustinian just war paradigm, did America ensure that this war would conclude with a lasting peace? The answer is a tentative "yes." Germany was eventually reunited. Even before the end of the Cold War, West Germany quickly recovered from the war and regained its economic strength. Apart from wars in the Balkans and recent events in Ukraine, Europe has been at peace since 1945. Japan rebuilt itself from almost the ground level and has become one of the world's safest and most prosperous nations. It, too, has been at peace since having virtually disbanded its military forces. Of course, the Second World War ended with a Cold War that contained the prospect of a ruinous conflict with the Soviet Union. It is likely, though, that America would have been unable to prevent this dynamic.

Looking forward after this period, as a nominally Christian nation that had done so much to preserve stability in the world, the question would be whether America could continue to apply *jus post bellum* efforts such as these elsewhere. During the Cold War, America would not have the same opportunity to replicate the successes that it saw at the end of the Second World War. That opportunity, though, would present itself again after the invasion of Iraq.

The American Experience in Iraq

Shortly after the turn of the twenty-first century, the United States invaded Iraq. There were many causes. Though Iraq's leader, Saddam Hussein, was not directly involved in the terrorist attacks of September 2001, there was a sense that Iraq was a destabilizing force in the region.[259] Iraq had carried on a war with Iran in the 1980s and it had invaded Kuwait in 1990. Following its defeat at the end of the First Gulf War, Iraq was placed under sanctions, to include a no-fly zone and frequent inspections of its nuclear program.

The nuclear program became an issue as did the disclosure of Iraq's chemical and biological weapons capabilities. Based on all these factors, the United States sought to invade Iraq, with the aid of its allies, and replace Saddam Hussein with a more democratic regime. The actual invasion concluded by April 2003 and Saddam Hussein was captured later that year. The occupation, insurgency, and reconstruction efforts, however, lasted until a formal withdrawal in 2011.[260] That being said, elements of the American military remain in Iraq to the present day under the invitation of the host government.

Iraq's new government has remained in power, though its grip has always been tenuous.[261] A consistent issue throughout this portion of the study will be a lack of unity among the Iraqi people. This issue goes back to the initial formation of the country after the First World War. Iraq has always contained primarily Shiite Arabs in the south, Sunni Arabs near its capital, and a very different ethnic group, the Kurds, in the north.[262] The differing branches of Islam had been rivals for over a millennium and the Kurds actually desired independence. Before 2003, these groups were primarily unified by force. With Saddam Hussein being out of power, all *jus post bellum*

---

[259] David J. Lorenzo, *War and American Foreign Policy: Justifications of Major Military Actions in the US* (Taipei: Palgrave Macmillan, 2021), 239.
[260] Lorenzo, *American Foreign Policy*, 220.
[261] Allawi, *The Occupation of Iraq*, 453.
[262] Ibid., 19-21.

efforts had to account for these factors in addition to the fact that nearby Iran was more than willing to promote an insurgency inside southern Iraq.[263]

War Crimes and Accountability

One of the main goals of the war was to hold Saddam Hussein and those in his regime accountable for war crimes for the manner in which they treated their countrymen. Two crimes, in particular, are relevant. In one case, he was accused of ordering the torture or execution of thousands of inhabitants of the village of Dujail following a failed assassination attempt.[264] In another, he was accused of ordering the deployment of nerve and mustard gas against Kurdish villages in northern Iraq.[265]

At first, this did not necessarily mean that another war crime tribunal would take place. In fact, it was possible, for a time, that a regime change would have been acceptable. As Bass writes, "Before the Iraq war, Donald Rumsfeld, the U.S. secretary of defense, floated the idea of exiling Saddam and other top Ba'athists, with *de facto* impunity from war crimes prosecutions as a 'fair trade to avoid a war'."[266] The idea was that an invasion, and the subsequent occupation, could have been avoided if Saddam were replaced beforehand. Naturally, this would have meant that Saddam and his elites would have escaped justice. Bass notes, "This would have meant selling out Saddam's forthcoming trial, but Rumsfeld's suggestion was something that U.S. and Iraqi soldiers would presumably have agreed upon."[267] The calculus was that the transition to a new, presumably more enlightened, government would have been supported by the Iraqi army, and Iraq would have remained stable afterward.

The moral case for trying Saddam as a war criminal was relatively straightforward. Saddam did not make any serious attempts to hide his activities. As Patterson writes, "Hussein modeled himself on Joseph Stalin, so it should not be surprising that his reign was notorious for heavy-handed dealings against his own population, including torture, extra-judicial killing, and the use of chemical weapons on Kurdish villages, as well as attacks on neighboring countries."[268] Many of his acts would not become part of the eventual prosecution. Notably, he was the aggressor during the Iran-Iraq War

---

[263] Ibid.,184-185.
[264] Ibid., 36.
[265] Ibid., 41.
[266] Bass, "Jus Post Bellum," 405.
[267] Ibid.
[268] Patterson, *Ending Wars Well*, 78.

in the 1980s and during the invasion of Kuwait. He had also supported a harsh reprisal against the Marsh Arabs in southern Iraq following a failed rebellion in 1991.

The actual trial process took an interesting turn. When Saddam was first captured, an Iraqi Special Tribunal (IST), operating under the Coalition Provisional Authority tried him and some of his senior aides for "genocide, war crimes, and crimes against humanity."[269] Unlike the Nuremberg trials described earlier in this work, there was a considerable amount of time between arrest and trial, perhaps giving sufficient time to prepare evidence and an adequate defense.

In the intervening time before the trial, the Iraqi government gained a higher degree of autonomy and established a new set of judiciary processes. As Patterson writes, "By the time of Hussein's October 2005 trial, the Iraqi government had passed Iraqi Law Number 10 (2005), disbanding the IST and establishing the Supreme Iraqi Criminal Tribunal (SICT). The law set out the rights of the accused, procedures for judicial tribunals, the appellate process, and the like."[270] This change was important because it had two important implications for *jus post bellum* war crime trials.

First, by changing the judiciary body itself but still continuing the prosecution, the Iraqi government was essentially trying its own leader for war crimes. This was, perhaps, the first time that a UN nation had autonomously prosecuted its own leader.[271] Recalling earlier twentieth-century cases, the tribunals after the Second World War were conducted by occupying forces. In this case, the original plan was for a court composed of Iraqi judges, but under a Coalition occupation government, to conduct the prosecution. However, the Coalition Provision Authority transferred control back to the Iraqi government in 2004.[272]

Second, the new Iraqi government was responsible for choosing what crimes to prosecute and for carrying out the sentences. Saddam's crimes were well known to the Iraqi people, but he actually attended two trials, one for attacking the people of Dujail after the assassination attempt and one for military aggression and using poison gas against the Kurds.[273] This means that Saddam was held accountable, having received the death penalty, but

---

[269] Ibid., 83.
[270] Ibid., 83.
[271] Ibid., 84.
[272] Allawi, *The Occupation of Iraq*, 280.
[273] Patterson, *Ending Wars Well*, 84.

some groups, such as the Shiites of southern Iraq, never received justice. Of additional note, the new Iraqi government also tried a number of co-defendants, many of whom were executed or imprisoned.[274]

In summary, the *jus post bellum* accountability effort had mixed success, but it contained notable developments. Though Saddam was not tried for all his alleged crimes against humanity, it is interesting that the proceedings were conducted by a legitimate Iraqi government. In terms of world opinion, not every government supported Saddam's death sentence or method of execution. To be fair, though, the defendants who were executed after the Nuremberg and Tokyo trials received similar punishment. To put these issues into an Augustinian just war context, Saddam's removal and punishment were justified and acceptable. Also, given that capital punishment was common enough in Augustine's era, he may not have disapproved of Saddam's execution. He writes, "The same divine law which forbids the killing of a human being allows certain exceptions, as when God authorizes killing by a general law or when He gives an explicit commission to an individual for a limited time." [275] Using poisonous gas among one's own population and violent reprisals against innocents are crimes that merit capital punishment under many systems of human law.

Economic Reconstruction

Following the Coalition invasion, Iraq's economy was in serious trouble, though not necessarily due to the fighting. The invasion itself proceeded quickly. The problem was that Iraq had been at war from 1980-1991 and then was subject to economic sanctions in the years leading to the Second Gulf War. Much of the infrastructure was intact, but the Iraqi currency was near collapse, meaning that it was increasingly difficult to trade in the world market.[276] For the people themselves, employment within the Iraqi economy often meant government service, which was subject to political whims, or some small-level business or agricultural work.[277] The Iraqi government, in turn, depended heavily on oil exports, which were sensitive to changes in oil prices.

---

[274] Ibid.
[275] Augustine, *The City of God: Books I-VII*, 53.
[276] Joseph Sassoon, "Iraq's Political Economy Post 2003: From Transition to Corruption," *International Journal of Contemporary Iraqi Studies* 10, no. 1-2 (March 2016), 18.
[277] Sassoon, "Iraq's Political Economy," 19

While aid was provided during this period under the oil-for-food program, the level of corruption in Saddam's government ensured that the people who needed assistance often did not receive it. In fact, both before and after Saddam's fall from power, endemic corruption has continuously stifled economic growth.[278] This became an important issue after the United States attempted to rebuild the Iraqi economy in 2003 but saw little progress.

The actual *jus post bellum* economic reconstruction efforts had concluded by 2004, though the United States continued to provide assistance afterward.[279] While the Iraqi economy only improved slowly, two notable successes are apparent. First, banking reform and the reissuance of a stable currency meant that free trade could once again be established. Iraq was self-sufficient and required overseas imports. A stable currency that was accepted in world markets meant that needed goods could be purchased.

Second, as a defensive measure, Coalition forces ensured that Iraqi oil continued to be available for export. This was a significant challenge as the country relied on oil pipelines to deliver crude oil to Turkey and to terminals in the Arabian Gulf.[280] These pipelines were vulnerable to attacks by insurgents but generally remained open due, in part, to Coalition security efforts. Therefore, by the time of the official Coalition withdrawal in 2011, the Iraqi economy was stable and appeared to be experiencing moderate growth, though this growth was tied to oil revenues.

This economic revitalization effort was not nearly as successful as hoped, however. Four key issues, which, in some ways, were present before the invasion stifled economic growth at the time and continue to do so until the present day. First, the economy was predominantly composed of state-run enterprises. Christopher Coyne and Adam Pelillo write, "In Iraq, the initial governing authority – the Coalition Provisional Authority (CPA) – reneged on economic reform regarding the privatization of state-owned enterprises after seeing how the initial privatization subsequently affected unemployment and the economy in general."[281] This is very important for a variety of reasons.

---

[278] Ibid.
[279] Ibid., 20.
[280] Ibid., 19.
[281] Christopher J. Coyne and Adam Pelillo, "Economic Reconstruction Amidst Conflict: Insights from Afghanistan and Iraq," *Defense and Peace Economics* 22, no. 6 (2011): 630.

One impediment to economic reform was the sheer number of stakeholders with competing interests. As Coyne and Pelillo noted, economic reforms needed to satisfy not only the Sunni, Shia, and Kurdish interests but also the various smaller local factions.[282] In Iraq, the Sunni minority had been in power for some time. Recalling the accounts of Saddam's crimes, his treatment of the Kurds and of the Shiite Arabs, whom he considered to be allies of Iran, was especially harsh. It would have been exceedingly difficult for those people groups to work together.

Another issue was that, during the period of Coalition economic reconstruction efforts, the military was responsible for implementing the necessary reforms. At first glance, this may seem ideal as, for some *jus post bellum* efforts, the military is ideally, manned, trained, and equipped for rebuilding and reconstruction projects. A state economy, though, is, perhaps too complex for any kind of centralized organization to manage. Coyne and Pelillo make this point when they find that any assumption that a military can engage in economic planning is inherently flawed.[283] They make an important point. Economies depend on interactions between citizens, in addition to government support. Furthermore, much depends on confidence, availability of capital, and supply of labor.

Any military, even one that is highly trained, may be able to rebuild infrastructure or industry and, thus, have a positive impact on the economy. Anything more than that would rely heavily on popular involvement. In comparing the mixed success of Iraqi economic revitalization with the much higher degrees of post-Second World War successes, one clear difference is the lack of popular support. Due to sectarian differences and a feeling that the Coalition-led effort lacked legitimacy, economic development never truly received enthusiastic nationwide involvement. As Bass writes, "There can be debates about how much economic responsibility lies with the victors and how much with the vanquished, but one must insist on stringent requirements of consent and responsibility to the Iraqi people."[284] If the people themselves, who are the prime stakeholders of economic activity, are unwilling to reform corrupt practices, protect infrastructure, and move away from government-controlled industry, then the results will continue to be sub-optimal.

An additional concern has been the lack of secure property rights. Even during Saddam's era, building and maintaining housing and businesses

---

[282] Ibid., 630-631.
[283] Coyne and Pelillo, "Economic Reconstruction," 634.
[284] Bass, "Jus Post Bellum," 408.

has been a challenge. As Coyne and Pelillo note, "If political institutions are characterized by corruption and unchecked power, this will stifle economic activity because property rights may be insufficiently protected or regulatory hurdles (including bribes) to exchange, investment, and innovation may be too significant for local citizens or firms to overcome."[285] Before the invasion, the need to use bribes and remain loyal to the correct political party was a necessity. Afterward, with the rise of militia groups who could often maintain local control and seize property for themselves, maintaining any investment became difficult to achieve. As a result, economic activity, especially outside the major urban centers, remained at a low level.

Some notable lessons were learned during this period of reconstruction. During the post-war reconstruction of Europe, funding was applied inefficiently, but the end results were positive because sufficient resources were present and because the nations, in general, seemed to work well together. In an even more challenging environment, such as Iraq, it becomes apparent that reconstruction efforts must take place under a very high degree of oversight.[286] Fortunately, Iraq retained a wealth of natural resources that could be harvested and shipped out of the country. However, given that the oil industry was a state-run enterprise, checks and balances needed to be in place to ensure that the oil profits went where they were needed. As Bass writes, "There must be effective structures of oversight and shared economic decision making in place – all the more so if some of the resources are Iraqi, as will be the case if Iraqi oil is used to help with the restoration."[287] Unlike post-war Europe and Japan, Iraq had what it needed to begin rebuilding. The question was how to use those resources wisely.

Coalition forces also needed to become more judicious in their use of aid. Since a conflict was still occurring during the time of reconstruction, the issue was not simply about wasting economic aid. It was important that the economic aid did not, in fact, support the enemy. Coyne and Pelillo discuss a failure in this calculus when they write, "Specifically, consider the $644 million 'Community Stabilization Program' (CSP) in Iraq, which was suspended due to significant fraud and waste. An audit by the USAID's Inspector General found that some of the funds allocated for weakening the insurgency actually went to the insurgents."[288] In some ways, this was similar to the humanitarian aid situation in post-war Europe in which local politics ensured that goods were not distributed equitably. To add additional

---

[285] Coyne and Pelillo, "Economic Reconstruction," 636.
[286] Ibid., 640.
[287] Bass, "Jus Post Bellum," 408.
[288] Coyne and Pelillo, "Economic Reconstruction," 640.

complexity, in Iraq, local politics often meant that some stakeholders were continuing to support enemy forces.

This leads to, perhaps, the most important lesson of the Iraqi economic reconstruction effort. It is exceedingly difficult to rebuild before the war has concluded. The original thought process may have seemed sound as insurgencies are long-term affairs and popular support is needed to defeat them. Coyne and Pelillo argue, "Economic reconstruction has been viewed as part of a broader strategy to 'win hearts and minds' of domestic authorities, citizens, and insurgents in order to end conflict."[289] It is possible that this approach may have borne fruit. However, as discussed, economic planning is almost impossible with a very high degree of effort and popular support. An active insurgency made matters even more "highly complex and uncertain."[290] To put this another way, Coalition forces were actively competing against insurgencies for popular support.

The issue was not only that the persistent presence of low-intensity warfare created an overly complex political situation. Those who were charged with physically performing repairs to economic structures required protection and the infrastructure itself remained highly vulnerable. Coyne and Pelillo state, "Insurgents often target aid workers as well as infrastructure projects that have been administered by foreign reconstruction authorities. In Iraq, for instance, insurgents often targeted oil pipelines and other major infrastructure programs administered by the USA and its allies."[291] As a result, Coalition forces were required to protect aid workers, protect infrastructure, earn popular support, and defeat insurgents simultaneously.

In summary, though, the economic reconstruction of Iraq could still count as a success, though certainly not as successful as previous similar *jus post bellum* attempts. Following the official departure of Coalition forces, the Iraqi economy has remained relatively stable, and the new monetary system seems to have worked well. Aid is still provided, to an extent, but the Iraqi economy is, essentially, sustaining itself. Recent conflicts in northern Iraq with the Islamic State did challenge this dynamic, but that issue seems to have abated for now. In terms of how this reconstruction effort was in alignment with Augustinian just war principles, it may be fair to say, tentatively, that it did help ensure a lasting peace. Iraq has not been free from war since 2011,

---

[289] Ibid., 627-628.
[290] Ibid.
[291] Ibid., 628.

but concerns about the nation being a destabilizing force in the region have abated.

Political Transformation

Though Saddam Hussein was removed from power within a short period after the invasion, the challenge afterward was to set up a new, functioning government, deal with the Iraqi military, and institute liberal reforms. The new government has had to deal with the factionalism described earlier in this work as well as insurgencies and a new war with the Islamic State, but it remains functional, if divided by a host of political parties and factions.

Saddam's own party, which was a faction of the Ba'ath Party, was outlawed immediately after Baghdad was captured along with many of the stakeholders that it supported. As Patterson writes, "Paul Bremer disbanded the already hollow Iraqi Army, dismantled several government agencies, and barred Ba'athist Party members from office."[292] Briefly, the Iraqi Ba'ath Party was one element of a much larger socialist, pan-Arabic, nationalistic party that had regional organizations throughout the Middle East.[293] It believed in central economic planning, thus explaining the state of Iraq's economy, and envisioned the formation of a strong, unified Arab state.

The Iraqi army, which was reputed to be highly capable before the invasion, was, for the most part, disbanded after the conventional conflict had concluded. Iraqi military forces were eventually reconstituted, but the new senior leadership positions tended to be filled with officers who were not loyal to Saddam but had strong ties with the new Iraqi government.[294] On a similar, note, with the Sunni members of the Ba'ath Party being out of power, formerly marginalized groups, such as the Shiites of southern Iraq, began to assert their dominance.

Though it may have been the correct decision to remove Iraq's Ba'ath Party from power, this policy came with significant consequences. Many of the nation's most productive citizens were suddenly unemployed. As Patterson notes, "Thus 'punishment' of the regime purged the government of all Ba'athist loyalists, but also had the effect of destabilizing the country and pushing much of the educated workforce, including doctors, teachers, and lawyers, out of gainful employment and into the confidence of

---

[292] Patterson, *Ending Wars Well*, 39.
[293] Allawi, *The Occupation of Iraq*, 39-40.
[294] Orend, "*Jus Post Bellum*," 584.

insurgent elements."²⁹⁵ On a similar note, most members of the Iraqi military suddenly found themselves unemployed. As Orend writes, "Many critics of the US occupation of Iraq argue that a key decision which helped to spark the ongoing insurgency was the US choice to disband promptly the 400,000-strong Iraqi army...and then leave them to their own devices. Plans for employing these potentially dangerous men, providing them with opportunities, should have been developed." As a result of these factors, it is easy to explain why a strong insurgency developed shortly after the war's conclusion.

In terms of *jus post bellum* development, there are many lessons to be learned from the U.S. occupation of Iraq. In hindsight, it is apparent that creating a large number of disenfranchised citizens can have negative consequences. To take a philosophical view, though, it is possible that the Coalition went too far in its desire to transform the Iraqi government into a stable democracy. Put simply, Iraq may not have been ready for democracy. Walzer essentially makes this argument when he asserts that the Coalition should have stopped at installing a government chosen and supported by the people but, instead, went too far and installed a government that it preferred.²⁹⁶ To put this another way, installing a new government to replace Saddam may have been wise, but perhaps given Iraq's history, a more authoritarian form of government should have been chosen.

Other just war theorists claim that political transformation should not have been attempted at all. Bellamy writes, "Once the threat of WMD Iraqi proliferation had been removed and reparations for damages caused by Iraq had been extracted, then the minimalist requirements of jus post bellum would have demanded a withdrawal from Iraq."²⁹⁷ This is an interesting point and, indeed, it may make sense to argue that a foreign government should not be responsible for imposing any form of government on a sovereign nation. However, this, too, may have created another leadership vacuum, thus hindering the goal of a lasting peace.

These negative views of Iraq's political potential in a *jus post bellum* environment may seem to be at odds with Augustine's views on war. However, it is important that modern scholars do not retroactively assign present-day values to Augustine. While a republic or a parliamentary democracy may seem ideal in the context of Western thought, Augustine experienced rule under an authoritarian emperor. In fact, Rome was an

---

[295] Patterson, *Ending Wars Well*, 83.
[296] Walzer, *Arguing about War*, 164.
[297] Bellamy, "The Responsibilities of Victory," 621.

empire because it had failed as a republic. As noted earlier in this work, he did not seem to have a problem with this type of government. Indeed, he advocated for enlightened Christian rulership supported by an active Christian populace. For Augustine, the goals were peace and stability so that the earthly city could be as close an approximation to the heavenly city as possible. With this in mind, he would have preferred an authoritarian form of government over an unstable democracy provided that the desired results could be achieved.

Humanitarian Aid

Though the conventional war in Iraq was brief, the need to provide humanitarian aid afterward soon became obvious. With the conclusion of formal hostilities and the rise of low-level conflict, much of the nation descended into civil disorder. As Patterson writes, "Perhaps the first decisive blow to hopes for a new Iraq was the experience of ordinary Iraqis themselves – broadcast worldwide – as law and order dissolved: widespread looting, criminality, revenge killings, and lawlessness were ubiquitous"[298] It could be argued that many of the sectarian divisions that had been kept in check while Saddam was in power were suddenly released due to the absence of central Iraqi authority.

Still, there is evidence that the need for humanitarian aid was not a surprise to Coalition forces. During this conflict, humanitarian aid was an integral part of the rebuilding effort.[299]

As discussed throughout this study, the delivery of humanitarian aid has been a complex process, often involving military, governmental, and non-governmental organizations, each bringing its own capabilities and each operating under different guidelines. This complexity became apparent to planners even before the commencement of *jus post bellum* activities.

Even before the war began, the American military assumed authority over all humanitarian efforts. As Sarah Lischer writes, "The Department of Defense established an Office for Reconstruction and Humanitarian Assistance (ORHA) in the Pentagon. The measure signaled a departure from established practice, in which the Department of State and the US Agency for International Development (USAID) oversaw humanitarian assistance

---
[298] Patterson, *Ending Wars Well*, 39.
[299] Sarah Kenyon Lischer, "Military Intervention and the Humanitarian 'Force Multiplier'," *Global Governance* 13 (2007): 100-101.

and economic development programs."[300] This shift appears appropriate for two reasons. First, the plan was to deliver humanitarian aid before the complete cessation of hostilities, meaning that the military would be responsible for security. Second, the delivery of humanitarian aid was a component of establishing military control throughout the county.[301]

Non-governmental organizations were allowed to operate within Iraq. However, they, too, were required to submit to American military authority. As Lischer writes, "Aid organizations were required to register with the Humanitarian Operations Center set up by the US government in Kuwait."[302] Again, the issue was that the American military was required to provide some level of security for these humanitarian aid workers. It was this issue, though, that provided the greatest challenge to American humanitarian aid efforts.

Put simply, in a country as large as Iraq, it is exceedingly difficult to provide nation-level security, even with a large military presence. In addition, there may be evidence that the American military placed a lower value on protecting NGO relief efforts.[303] These issues became apparent relatively quickly after the conflict began. As Lischer writes, "Attacks against aid workers in Iraq, including abduction and killings, have effectively curtailed the activities of NGOs and the UN. The International Committee of the Red Cross withdrew all its personnel following the October 2003 bombing of its Baghdad headquarters."[304] Therefore, it is apparent that the presence of sustained conflict undermined *jus post bellum* humanitarian efforts. Whereas the challenge after the Second World War was the efficient delivery of humanitarian aid, the difficulty in Iraq was the safe delivery of needed food and medical supplies.

In summary, there were several lessons to be learned from *jus post bellum* humanitarian efforts in Iraq. It could be argued that the desire to centralize the various aid organizations was a step in the right direction. Even if mistrust and communication issues hindered the overall effort, it is apparent that attempts were made to streamline the delivery of needed supplies and to provide security for aid workers. Some scholars have argued that humanitarian aid efforts were a complete failure due to a lack of

---

[300] Ibid., 105.
[301] Ibid.
[302] Lischer, "Military Intervention," 105-106.
[303] Allawi, *The Occupation of Iraq*, 124.
[304] Lischer, "Military Intervention," 106.

planning.[305] However, as discussed, these efforts were planned well in advance. Instead, reconstruction and humanitarian efforts should have waited until the conflict was finally concluded. Lischer makes this argument when she writes, "If even US military forces cannot perform humanitarian tasks adequately while under constant threat of attack, certainly aid workers cannot. No matter who provides humanitarian assistance to a needy population – soldiers or aid workers – security is a prerequisite for success."[306] Adequate security could only have been maintained after each element of the insurgency was defeated.

From an Augustinian just war perspective, the issue of humanitarian aid in Iraq presents a challenging question. How does a nation meet immediate humanitarian needs before the war has finally concluded? On one hand, the righteous nation has a responsibility to provide comfort to those who suffer due to war. However, the nation has an equally serious responsibility to protect its own citizens. Perhaps, the only ideal solution was the one proposed earlier in this work, which was to quickly support the transition to a form of government that would have been acceptable to the populace. Though delivering humanitarian aid through a friendly authoritarian regime may not have been ideal, or even acceptable in a Western context, this action may have been in better alignment with Augustine's goals of alleviating suffering and returning a nation to a state of peace.

Impact on Jus Post Bellum Development

It has become apparent that, after the conflict in Iraq had ended, the *jus post bellum* framework required a significant amount of revision. Before 2003, there was, perhaps, a sense that *jus in bello* and *jus post bellum* activities could occur simultaneously. In retrospect, it seems obvious that economic reconstruction and humanitarian aid must wait until the conclusion of hostilities simply because the personal security requirements would be too high, otherwise. It is true that some *jus post bellum* activities did take place before the Second World War had concluded. However, these activities tended to take place in secure, peaceful areas. In the case of Iraq, the presence of an active insurgency ensured that no area was truly safe.

Opinions vary regarding the cause of this failure. Iasiello finds that the United States simply did not adequately plan for the occupation after the conclusion of hostilities when he writes, "As recent events in Afghanistan and Iraq attest, nations must fight wars with a war-termination vision and

---

[305] Bellamy, "The Responsibilities of Victory," 602.
[306] Lischer, "Military Intervention," 107.

plan carefully for the post-conflict phase."[307] In hindsight, this point of view makes logical sense, but it is important to recall, that many activities, such as the consolidation of humanitarian work, began well before the war commenced. Patterson makes a similar, more generous, argument when he notes, "In fact, planning resulted in some successes, such as preserving Iraq's oil infrastructure."[308] This is a salient point. While humanitarian efforts were severely hampered, at least Iraq's economic lifeline remained intact. Keeping all points of view in mind, it may be most fair to say that the United States went into the war with some faulty assumptions about its ability to provide security nationwide.

There were some *jus post bellum* successes. Saddam Hussein had truly committed horrendous crimes and he was held accountable. As Patterson writes, "Justice, in this case the execution of Saddam Hussein, was punishment directed at the perpetrator for causing conflict and for violations of the war convention, and thus was narrow, targeted, and ethically sound."[309] As discussed previously, it is noteworthy that the new Iraqi government took it upon itself to try Saddam for war crimes. However, other efforts at political and economic reformation, such as the expulsion of the Ba'ath Party and the immediate drawdown of the Iraqi army seem to have been poorly planned. As Coyne and Pelillo write, some failures may be "because of inappropriate or irrelevant strategies and policies underpinning the initial occupation and reconstruction."[310] While some good did come from *jus post bellum* efforts in Iraq and some failures are apparent only in hindsight, the disruption created when so many former professionals and soldiers were suddenly and forcibly unemployed indicates a failure on the part of planners to fully understand the situation on the ground.

In summary, though, did the United States fulfill its original purpose for the war, and did it establish a lasting peace? The answer is a tentative "yes," but this peace is by no means stable, and, in fact, it remains to be seen whether it will be lasting. The purpose of the war, which was to create stability in the region, was somewhat divorced from the necessities that became apparent after the conclusion of hostilities.[311] The original reason for commencing a war in Iraq was to remove the threat of weapons of mass destruction and to hold Saddam Hussein accountable. It appears that this aim has been achieved. The inadvertent creation of a series of insurgencies made

---

[307] Iasiello, "*Jus Post Bellum*," 39.
[308] Patterson, *Ending Wars Well*, 2.
[309] Patterson, *Ending Wars Well*, 85.
[310] Coyne and Pelillo, "Economic Reconstruction," 627-628.
[311] Pattison, "Jus Post Bellum," 653.

the *jus post bellum* environment much more complicated but, by all accounts, the new government of Iraq remains in power and appears somewhat able to maintain order. *Jus post bellum* efforts in Iraq did not reproduce the successes of Germany and Japan and, in fact, were often poorly applied, but there is evidence that they may have worked. Only time will tell if Iraq's long-term stability is viable.

Comparison with Augustinian Thought

Would Augustine have approved of *jus post bellum* efforts in Iraq? It would be best to argue that he would not have approved of them because he would not have believed that the war itself was necessary. The issue is that the United States did not exhaust all possibilities before choosing to go to war and participated enthusiastically rather than warily. As Peter Lee writes, "Augustine's ideals were used in a pro-war capacity to support a case for intervention in Iraq that was proving increasingly difficult to substantiate in the current global polity, despite the accuracy of the claims that Saddam's evil actions had resulted in the suffering of his own people over many years."[312] To put this another way, though it was obvious that Saddam was oppressing his people, that may not have been enough of a reason to remove him from power.

A possible counter to this argument would be that a righteous nation has a responsibility to preserve peace so long as it has the ability to intervene. Indeed, this misunderstanding of the Augustinian just war theory may be partially to blame for America's desire to commence the war.[313] While Augustine certainly advocated for intervention in the interests of peace, he realistically realized that national resources are limited. As Lee writes, "Augustine did not try to avoid the difficulties of dealing with the contradictions involved in the Christian's encounter with the decidedly ungodly Earthly City."[314] Just as Christians cannot possibly cure all of the problems in their sphere of influence by themselves but must, instead, use discretion, rulers must be equally judicious with the resources at their disposal.

Two of the difficulties that presented themselves during the occupation of Iraq highlight this point. First, though Saddam Hussein was an unjust ruler, the correct action if war were necessary, in Augustinian terms,

---

[312] Peter Lee, "Selective Memory: Augustine and Contemporary Just War Discourse," *Scottish Journal of Theology* 65, no. 3 (July 2012): 322.
[313] Ibid.
[314] Ibid., 309.

would have been to replace him with a person that the populace could support. Augustine would have approved of a righteous, yet authoritarian, ruler but this would have been unacceptable in a Western context. Second, with the understanding that the Roman military was limited in size in Augustine's day, it would be reasonable to assume that he would not have advocated for simultaneous war and rebuilding efforts.

At this juncture, it may also be relevant to note a significant shift toward an Augustinian just war construct that took place shortly before the conflict in Iraq. Through the Vietnam War, the United States had relied on the conscription of necessary military personnel through a national draft system. President Nixon abolished the draft in 1973 and, since that time, the United States has relied upon an all-volunteer military.[315] The challenge, of course, was that the size of an all-volunteer force would necessarily be limited. However, this concept appears to fall very much in line with Augustine's view of the citizen-soldier.

Of additional note, there is evidence that one major component of the Augustinian just war construct was left unattended. In Iraq, Christians faced a much higher degree of persecution after Saddam Hussein was replaced. According to Shak Hanish, "They have been the target of extremist Islamic groups – whether Sunni or Shiite – to eliminate them physically or to destroy their places of worship, shops, and even their homes."[316] The implication is that Coalition forces did not create safe spaces for the spiritual development of minority faiths, to include Christianity. When comparing the whole of the *jus post bellum* efforts in Iraq to their supposedly Augustinian origins, this is a significant problem.

The American Experience in Afghanistan

The conflict in Afghanistan represents, perhaps, a low point in *jus post bellum* development and in the application of Augustinian just war practices. The war began in 2001 with the rapid removal of Afghanistan's governing party, the Taliban, from power. *Jus post bellum* efforts continued for almost twenty years but were hindered by an ongoing series of insurgencies, internal corruption, and a lack of respected centralized authority.[317] When the

---

[315] Beth Bailey, *America's Army: Making the All-Volunteer Force* (Cambridge, MA: Harvard University Press, 2009), 3.
[316] Shak Hanish, "Christians, Yazidis, and Mandaeans in Iraq: A Survival Issue," *Domes* 18, no. 1 (Spring 2009), 3.
[317] Thomas P. Cavanna, *Hubris, Self-Interest, and America's Failed War in Afghanistan: The Self-Sustaining Overreach* (London, UK: Lexington Books, 2015), 114.

United States did finally depart in 2021, the Taliban had reasserted control over the county before the withdrawal was even complete.

The true reason for commencing the war was to hold the terrorists who were responsible for 9/11 accountable, but the mission quickly transformed into a desire to ensure that Afghanistan could never again be a state that would support further acts of terrorism.[318] By necessity, the expanded mission required the *jus post bellum* activities that would lead to a lasting peace. Also, of note in regard to the development of just war theory, all of NATO joined a war in response to an attack on one of its member nations.[319]

History and geography are defining factors in the failure to create a safe, stable democratic nation in this region of the world. Whereas the situation in Iraq involved dealing with approximately three people groups and their associated subgroups, Afghanistan has always been composed of a patchwork of tribes. The Tajiks, for example, formed the governing body of the nation, but this tribe was actually a minority.[320] To use another example, the Pashtuns, who have been a majority people group, have a tribal area that extends deep into Pakistan.[321] To make matters more challenging, the mountainous terrain of the county complicates travel and communication, and, in fact, tends to contribute to the isolated nature of many of Afghanistan's people groups. As a result, tribal loyalties have tended to matter more than any loyalty to the central government in Kabul.

War Crime Trials and Accountability

The war in Afghanistan began as an effort to hold the architect of the 9/11 attacks and his supporters accountable for their actions. Though the nature of the war changed, the idea of bringing terrorists to justice remained. The actions that took place in support of this idea and the thought processes behind them had profound effects on just war theory and on Augustinian-related thought.

It is vitally important to understand that the legal system within Afghanistan is almost unique. While, technically, there has always been a state judicial system, justice has been predominantly a tribal matter, with many

---

[318] Ibid., 42.
[319] Cavanna, *America's Failed War*, 42.
[320] Ibid., 27.
[321] Ibid.

decisions being made during a *jirga* and in accordance with Islamic law.[322] In fact, some of the main reasons why Osama bin Laden was not held accountable until 2011 were that he had taken refuge in a Pashtun-held area and, under the tribal code (*Pashtunwali*), he was able to claim asylum. He was then able to flee to Pashtun tribal lands across the border in Pakistan.

This decentralized system of justice has presented both problems and opportunities for Afghanistan's occupiers. Before 2001, the Taliban won a great deal of respect due to their ability to maintain order nationwide. As Cavanna notes, "The Taliban also won popular support thanks to their determination to punish the warlords' abuses, and to offer a brutal yet effective justice system."[323] While it may have been theoretically possible for NATO to establish a central legal authority to prosecute war criminals, a critical factor was that Afghans only tended to trust those who applied Islamic law.[324]

Therefore, while there was a judiciary body present in Kabul, many of the *jus post bellum* efforts designed to hold war criminals accountable either took the form of kinetic action, such as in the case when Osama bin Laden was killed or of a relatively novel judicial process. After the war in Afghanistan began, suspected war criminals were captured and sent to an American military base in Cuba. As Charles Smith writes, "Two months after the order by President Bush that provided for military commission, the first 774 prisoners arrived in Guantanamo Bay, Cuba."[325] This prison would continue to hold suspected terrorists until the present day.

The prisoners taken to Guantanamo did not have the same rights as, for example, the prisoners who were present during the Nuremberg tribunal. First, through a concept known as "anticipatory justice," the United States was able to hold suspected terrorists who may not yet have committed an act of terrorism.[326] Furthermore, when tried, the burden of proof required for a conviction was much lower than it would have been under the typical

---

[322] Michael Newton, "Community-Based Accountability in Afghanistan: Recommendations to Balance the Interests of Justice" in *Jus Post Bellum and Transitional Justice*, eds. Larry May and Elizabeth Edenberg (Cambridge, UK: Cambridge University Press, 2013), 75.
[323] Cavanna, *America's Failed War*, 23.
[324] Ibid., 117.
[325] Charles Anthony Smith, *The Rise and Fall of War Crimes Trials: From Charles I to Bush II* (Cambridge, UK: Cambridge University Press, 2012), 249.
[326] Ibid., 244-245.

American legal system.[327] While this modified legal system may seem unusual, it was derived from the precepts of the Geneva Convention which, as noted earlier in this study, was based, in part, on Augustinian principles. The Geneva Convention was careful to delineate lawful and unlawful combatants during a conflict. It was expected that lawful combatants would be treated humanely. Terrorists, however, were considered to be unlawful combatants and, as a result, could not expect to enjoy the protections of the Geneva Convention.[328] This was an interesting turn in the history of *jus post bellum* justice as, due to their designation as unlawful combatants, suspected terrorists were afforded fewer protections than the suspected war criminals in Nuremberg and Tokyo.

This dynamic led to another issue that hampered any attempts to deliver effective postwar justice. In an effort to capture suspected terrorists in hiding, the United States offered cash incentives for information leading to a successful apprehension. As Smith writes, "Many of the detainees were taken into custody as a result of the $5000 bounty offered by the United States for anyone allegedly involved with the Taliban, Al Qaida, or terrorism in general."[329] Of course, this would create an incentive for people to be accused of terrorism based on very little evidence.

In summary, the *jus post bellum* legal environment in Afghanistan was incredibly complicated due to the lack of any type of respected central authority. As Newton writes, "Theoretical constructs of *jus post bellum* are most challenged by the realities of a complex choreography of authoritarian local actors capable of administering justice/reconciliation grounded in sociological legitimacy."[330] American efforts to craft a more effective system through a reinterpretation of the Geneva Convention may have done more harm than good. To put these issues in an Augustinian perspective, the moral structures designed to protect the innocent from conflict should not have been reinterpreted to strip suspects of their legal rights. It may have been more in keeping with the Augustinian spirit to afford suspected criminals, at a minimum, normal legal protection. With this thought in mind, it may be possible to compare the treatment of suspected terrorists with the Roman Empire's treatment of the Donatists. However, while Augustine may have approved of their persecution, it is likely that he would have required a normal standard of proof of their heretical involvement beforehand. As Scalise notes, Donatists were to be persecuted through the use of secular laws

---

[327] Ibid.
[328] Ibid., 250.
[329] Ibid., 251.
[330] Newton, "Community-Based Accountability," 74.

and out of a loving sense that they would repent.[331] Though the Donatists certainly did not enjoy the legal protections that criminal defendants would have in later days, there is a sense that their prosecution was fair by the standards of the day.

Economic Reconstruction

Afghanistan's economy, which was never truly robust, had been severely affected by decades of war going back to the initial Soviet invasion in 1979.[332] As a predominantly agricultural economy, the nation's requirements and dynamics were very different from those present in Germany, Japan, and even Iraq. In short, the vast majority of Afghans have lived at the subsistence level with very little hope of obtaining what would be considered a comfortable living by Western standards. Barfield writes, "The scarce arable land there produced little in the way of surplus food or cash crops. In such a subsistence economy, it was difficult to accumulate substantial wealth when what existed was consumed in meeting the obligations of hospitality and other expressions of generosity that maintained social status."[333] The concept of *Pashtunwali* actually worked against economic advancement as members of the majority tribe were required by custom to share resources with each other. Briefly, *Pashtunwali* is an ancient code of ethical behavior among Pashtuns that, notably, includes complex requirements for hospitality and asylum.[334]

One notable exception to this dynamic is that Afghanistan has historically been a major producer of one important cash crop: opium. At one point, in fact, Afghanistan "provided 90 percent of the world's illegal production."[335] The allowance or prohibition of this crop will become an important factor in America's *jus post bellum* economic development plans.

In general, reconstruction focused on those issues that could be corrected in the context of a predominantly agricultural society. Though these efforts were undoubtedly affected by the level of corruption in the country, there is evidence that they were quite successful. As Coyne and Pelillo write regarding the positive changes that took place during the American occupation, "The rebuilding of schools, hospitals, and infrastructure has taken place despite the presence of violent insurgencies.

---

[331] Scalise, "Exegetical Warrants," 500.
[332] Barfield, *Afghanistan*, 185.
[333] Ibid., 217.
[334] Ibid., 203-204.
[335] Ibid., 239.

School enrollment has skyrocketed in Afghanistan, and girls, who were previously unable to attend school under Taliban rule, now comprise a third of the country's education enrollment."[336] The idea was that these efforts would improve the quality of life of the populace, thus influencing popular support towards NATO forces and away from insurgent elements. NATO forces could never expect to transform Afghanistan into an economic power as had been accomplished in Germany and Japan, but tangible results were possible.

It is possible that this *jus post bellum* effort could have been fully successful despite the time and cost requirements. However, reconstruction efforts in Afghanistan were never consistent, despite the fact that they took place over the course of two decades. One reason is that the concurrent war in Iraq became a major focus for a period of time. As Barfield writes regarding this period, "The rural economy got no attention, and progress on infrastructure repairs and improvements moved at a glacial pace despite large outlays of money."[337] Even when Afghanistan was a serious focus, reconstruction policies changed from year to year, such as when opium cultivation would be acceptable at one point and subject to sanctions at another.[338] Furthermore, despite the fact that the United States remained in Afghanistan for such a long period of time, there was a sense among the populace that American involvement was temporary.[339]

One major implication for *jus post bellum* development as a result of reconstruction policies in Afghanistan is that they could have been successful if they had been consistent. Despite a common assertion that Afghanistan is "the graveyard of empires," there is evidence that previous occupation efforts have been popularly accepted provided that the needs of the people were taken seriously. For example, the founder of the Mughal dynasty was widely respected due to his "love of Kabul."[340] The lesson for jus war theorists is that economic reconstruction, though expensive and, at times, limited in scope, could work with the appropriate level of patience and persistence.

In terms of Augustinian just war thought, a major lesson is that a type of lasting peace or, at least, a higher degree of stability, can be achieved provided that the righteous nation is willing and able to work alongside the populace of the occupied nation. This work can be expensive and can, in fact,

---

[336] Coyne and Pelillo, "Economic Reconstruction," 628.
[337] Barfield, *Afghanistan*, 236.
[338] Coyne and Pelillo, "Economic Reconstruction," 630.
[339] Coyne and Pelillo, "Economic Reconstruction," 632.
[340] Barfield, *Afghanistan*, 263.

take decades. Furthermore, if a low-intensity conflict that continues after the conclusion of the war cannot be avoided, then a high level of security for reconstruction efforts will be required. However, if the human needs of the populace are met, then success, and the goodwill that results, can be possible.

Political Transformation

On a similar note, there are indications that it may have been possible to achieve a stable, respected Afghan government through *jus post bellum* efforts provided that the desires of the people were understood. Contrary to a common belief that Afghanistan is ungovernable, within relatively recent history, its monarchy enjoyed popular support.[341] Barfield addresses this issue when he writes, with great optimism, "What Afghanistan sorely lacks today are leaders who have the talent to move the country from war to peace and lay the foundations of a stable future."[342]

The central issue for any successful government is an understanding that the diverse tribes of the region prefer to have a high degree of autonomy and will not tolerate authoritarian rule.

After 9/11, the United States was focused on removing the Taliban from power, ending support for terrorist organizations, and finding Osama bin Laden. In close coordination with a coalition of Afghans, then known as the Northern Alliance, the United States achieved this aim but, at the time, did not have any long-term plans for reestablishing a central Afghan government.[343] Later, an interim government was established, and negotiations began that would lead to a permanent Western-style government.

Eventually, with the support of the United States, a *Loya Jirga*, or council of elders, created the new constitution and Hamid Karzai became the president, but Karzai's government worked to marginalize political factions in an attempt to centralize authority in Kabul.[344] To put this into context, Afghans had traditionally made significant decisions through group consensus. For larger decisions, such as the *Loya Jirga*, well-respected, elder representatives from various groups are brought together. Though Karzai's

---

[341] Safiullah Taye, "Afghanistan's Political Settlement Puzzle: The Impact of the Breakdown of Afghan Political parties to an Elite Polity System (2001-2021)," *Middle East Critique* 30, no. 4 (October 2021), 336.
[342] Barfield, *Afghanistan*, 263.
[343] Taye, "Afghanistan's Political Settlement," 338.
[344] Ibid., 341.

government had the appearance of a Western-style parliamentary government, this form of government was acceptable under the premise that political parties would be able to successfully represent their constituents' interests in Parliament.

It is important to note that the Afghan concept of a political party is somewhat different from a Western conception of the same term. Afghan parties have tended to be regional and, as a result, have represented the interests of a tribe and local area.[345] Since Afghanistan had been transformed into a democratic form of government, regular elections were to take place. This dynamic could have been successful. However, the Karzai government, with the intent of diluting the power of political parties, required citizens to vote for individuals rather than for parties, thus creating competition within each party.[346] The end result was that "political formation from 'party' politics to an elite polity fragmented the country even further, severely hampering efforts to build a functioning and peaceful democratic state."[347] To put this another way, elections tended to ensure that the same popular figures remained in Kabul while regional authorities were, themselves, highly divided. To make matters more challenging, there were widespread allegations of voter fraud, thus undermining the legitimacy of central authority.[348] Finally, as NATO support came to an end in 2021, the Taliban took advantage of this dynamic and quickly reasserted control over the country, beginning with the exterior provinces.

Though *jus post bellum* political efforts in Afghanistan ended, after twenty years, with the same party back in power, it may be fair to say that there were some positive developments. The use of a representative council of Afghans, in accordance with Afghan tradition, to select a mutually agreed upon form of government appears to have been a step towards creating a lasting peace in the region. As was the case in Iraq, the presence of an active insurgency certainly complicated *jus post bellum* efforts, but significant progress was made, even if that progress did not exactly match the Western notion of an effective democracy. It was the devolution of that that government that doomed the enterprise to failure. It may be best to say that NATO's fault was that it failed to intervene and interrupt this process before it failed completely. Of course, that solution may have led to even deeper problems.

---

[345] Ibid., 344.
[346] Ibid.
[347] Ibid.
[348] Ibid.

Was the political transformation of Afghanistan in accordance with Augustinian principles? The tentative answer for at least part of the effort is "yes," but a deeper commitment to a long-lasting and consistent effort would have been required, to include support for the people over support for an increasingly wayward central government. It is apparent that initial efforts to create a lasting peace bore fruit because the Afghan people were a priority. Contrary to some opinions, the Afghans themselves desired order and stability. As Barfield notes, "Afghans today see the need for security as a basis to build a better economy, a more stable society, and a brighter future for their children."[349] It was a focus on the welfare of the people that had the potential to create the first instance of lasting peace in decades and it was a lack of complete focus on the needs of the people that led to their mistrust. When NATO forces finally began to withdraw, leaving only an unpopular government in Kabul, that mistrust led to open support for the Taliban.

Humanitarian Aid

Humanitarian aid efforts in Afghanistan bore many resemblances to those in Iraq but were often focused on education, stabilizing the provinces, and bolstering the legitimacy of the government in Kabul.[350] Regarding the focus on education, as discussed previously, school attendance by Afghans, especially females, increased significantly between 2001 and the conclusion of *jus post bellum* activities. In some ways, this effort would appear to be a success, though it did have notable drawbacks. Conservative elements within the populace, to include the Taliban, viewed Western education as an attempt by occupiers to impose Western values.[351] This, of course, would have alienated some segments of the population and caused many to see the Kabul government being supportive of these measures.

Attempts to stabilize the provinces as part of a humanitarian mission did not succeed. As William Maley writes, Provincial Reconstruction Teams "too often became substitutes for the Afghan state, creating perverse incentives for local political actors. At the same time, however, they proved unequal to the near-impossible task of stabilizing Afghanistan on a province-by-province basis."[352] These teams, which were composed of diverse groups of military members, were intended to reach out to the populace at the local

---

[349] Barfield, *Afghanistan*, 262.
[350] Peter Marsden, *Afghanistan: Aid, Armies, and Empires* (London, UK: I.B. Tauris, 2009), 120.
[351] Ibid., 125.
[352] William Maley, *Transition in Afghanistan: Hope, Despair, and the Limits of Statebuilding* (London, UK: Routledge, 2018), 43.

level and provide modest infrastructure, security, and humanitarian aid support. While this may have seemed to be a worthwhile attempt to provide direct support, these teams became, essentially, foreign representatives of the Kabul government and, thus, could never truly integrate with the people they served.

In a way, Provincial Reconstruction Teams also became substitutes for non-governmental organizations. As was the case in Iraq, these organizations had difficulty operating in the presence of an ongoing insurgency due to security requirements. As Peter Marsden writes, "The combination of the resistance movement and the campaign of terrorism also made it increasingly difficult to deliver reconstruction assistance. Thus, the aid programs delivered by the UN and NGOs were progressively withdrawn from more and more areas or were reduced in scale."[353] Provincial Reconstruction Teams, though, could continue to operate because they contained their own security component.

Other support included attempts to bolster national forces such as the Afghan National Army and the Afghan National Police.[354] While this may seem to be far removed from the idea of providing direct aid to an affected populace, the idea was to help the government establish and maintain a much higher level of security outside the capital. This may have been a sensible *jus post bellum* construct given that stability could foster trust in the government and allow a higher degree of practical aid to flow to the provinces. While the Afghan provinces had always demanded a level of autonomy, it is possible that these forces, used locally for the purposes of establishing a sense of security, could have worked alongside local leaders.

All of these efforts, though, failed to produce lasting results as the needs of the populace were never truly met. To use one example, Marsden writes, "The Afghan refugee population in Pakistan in 2005, at 3 million, was no less than it had been in 1992 or 2002."[355] Distrust in the government remained high not only due to the level of corruption in the Karzai regime but because government officials themselves "contained many individuals whose human rights record was not significantly better than that of the Taliban."[356] Furthermore, the rapid collapse of governmental security entities following the departure of NATO forces made it clear that those

---

[353] Marsden, *Afghanistan*, 130.
[354] Ibid., 121.
[355] Ibid., 122.
[356] Ibid., 126.

organizations had never progressed to the point where they could support themselves without foreign aid.

As was the case in Iraq, one major *jus post bellum* issue as related to humanitarian aid was the absolute necessity of ending hostilities before the commencement of post-war activities. Attempts to circumvent this reality have met with failure due to security concerns or to a failure to fully engage with the populace. It may be true that, in nations such as Afghanistan, creating a fully secure environment may be exceedingly difficult. If so, then this issue must be a component of any pre-war planning process.

In terms of the Augustinian just war theory, the humanitarian aid effort in Afghanistan did not succeed because it failed to truly support the immediate material needs of the people. It is important to recall that Augustine was a pastor who genuinely advocated for the material and spiritual needs of the people under his care. Though the people of Afghanistan are certainly not from a Christian context, their needs for medical care, personal welfare, peace, and space for spiritual development remain the same. By focusing on larger-scale security concerns rather than individual material needs, NATO achieved neither personal welfare nor national security.

Impact on Jus Post Bellum Development

In some ways, the *jus post bellum* lessons learned in Afghanistan have mirrored the lessons learned in Iraq. This is not surprising considering the overlapping timeframes. For example, in Afghanistan, security concerns hindered the delivery of humanitarian aid to the point where international organizations, such as those that operated after the Second World War, essentially ceased to function. Efforts to mitigate this risk failed to completely overcome it.

Indeed, the common factor in both post-war redevelopment efforts was that the actual fighting had not concluded. Besides humanitarian aid efforts, this affected the Coalition's ability to improve infrastructure, earn the trust of the populace, and help the society form a legitimate government. Establishing a respected government was, perhaps, more of a problem in Afghanistan even though there were indications that the process had the opportunity to be successful. Even though finding a replacement government was a secondary consideration when the invasion in 2001 began, using the *Loya Jirga* as a means to find a popular replacement meant that the needs of the people were heard. The failure of this government came later, and this failure was exacerbated by the fact that NATO forces needed to balance security concerns against the increasing unpopularity of this government.

As a result of factors such as these, the entire enterprise failed in 2021 when NATO forces departed, the government quickly collapsed, and the same Taliban that had been defeated in 2001 returned to power. Comparing this failure to the tremendous successes that followed the Second World War, one must wonder where the future of *jus post bellum* lies. The four post-conflict elements described in this study will need to be re-evaluated before being attempted again in the future.

Regarding personal accountability and the prosecution of war crimes, there is a sense that this effort could continue into the future, but it would need to contain the elements of fairness and respect for the rights of the accused that were present in previous *jus post bellum* scenarios. Declaring all potential suspects as unlawful enemy combatants proved to be a tortured means of applying just war theory that, in the end, eroded the legitimacy of the process.

The other three pillars of *jus post bellum* were hampered by a similar problem. They competed with the security needs caused by the continual presence of ongoing conflict. The government had to be supported because it needed to be part of the fight against several active insurgencies. Furthermore, NATO policies in support of this government were constantly changing in response to these insurgencies, giving the appearance of instability. Humanitarian aid was limited because it could only be distributed in those areas that were deemed secure. Finally, infrastructure could have been improved, even if it was not to the level of a modern Western state, but the application of infrastructure improvement efforts was haphazard due to the level of corruption in the provinces and the competing requirements of supporting the people while hindering support to the very insurgencies that relied upon the people. The support and subsequent ban on opium cultivation highlight this point.

In summary, it is apparent that, in the future, *jus post bellum* efforts must take place after the fighting has stopped. It may be possible to provide post-war support in fully secured areas, as was done during the Second World War, but humanitarian needs and the practical needs of fighting a war cannot compete with each other. If there is to be a case where the environment does not make the application of post-war support possible, then that issue needs to be realized well before hostilities commence, and, to use an Augustinian view, the decision to go to war should be deliberated upon with great apprehension.

Comparison with Augustinian Thought

What would Augustine have thought about *jus post bellum* efforts in Afghanistan? It is very likely that he would have strongly disapproved of the entire enterprise. The problem is not that humanitarian intervention was inadequate or that the United States supported an increasingly corrupt, authoritarian central government, or even that the United States used Guantanamo Bay as a prison camp. Looking back at the reasons for beginning the war, under Augustinian thought, the United States should not have commenced a war in Afghanistan at all.

The righteous nation should only commence a war in order to prevent some type of greater level of suffering. Even then, the decision to commence a conflict should be taken very reluctantly. None of these criteria were met in 2001. While America had been attacked, thus initiating a NATO response, Afghanistan was not the belligerent power. Instead, the ruling party in Afghanistan harbored terrorists, causing the American President to declare the Taliban as a supporter of terrorists.[357] While the Taliban's actions may have been reprehensible, an invasion would not have been the correct response because the level of suffering caused by the invasion would have been much greater than any future suffering planned by the Taliban. Furthermore, the relevance of local tribal customs in harboring fugitives, to include the fact that suspected terrorists could find refuge in Pakistan, meant that it was overly simplistic to see the Taliban as a hostile government.

Furthermore, the war in Afghanistan was not conducted reluctantly. The speed at which it commenced is an indication of a lack of considered deliberation and, in fact, this work has already indicated that many *jus post bellum* activities were not planned until well after the invasion had commenced. America was certainly correct in its desire to hold the 9/11 perpetrators accountable for their actions, and, certainly, Augustine would have approved of bringing criminals to justice, but an invasion of a sovereign nation to find those criminals was a step too far. The fact that Osama bin Laden was eventually found in Pakistan highlights this point. Also of important note, the condition of the people of Afghanistan had very little to do with the invasion itself. While the Taliban would, unquestionably, have fit the definition of an unjust foreign ruler, the party was brought into power because it was superior to previous alternatives. The people were consulted

---

[357] Lee, "Augustine," 321.

in the formation of a new government, but, for the most part, *jus post bellum* activities were not focused on popular needs.

Augustine would also have disapproved of the use of an all-volunteer army to support these goals. Lee writes, "Augustine's concern for Christian leaders who embarked upon war and Christian soldiers who waged war was rooted in his wider concern that all Christians should pursue an ultimate *telos* located in the City of God."[358] When civilian volunteers join a military force to fight for a righteous cause, for the sake of the soldiers and the sake of the nation's leaders, the cause itself should be righteous. When the state and, by extension, its citizens, are caused to inflict harm on others, it should be clear to them that they are acting righteously. Granted, the overtly Christian elements of Augustine's views in this matter do not match twenty-first- century secular reality. However, it is also important to remember that Augustine himself was a realistic man. Even in his own day, it would have been impossible to say that Roman emperors, and those who served under them, were all devout believers. Even then, Augustine, as a pastor, was concerned about the moral welfare of the citizenry in his own day as he would surely be concerned about the spiritual condition of today's citizen soldiers.

---

[358] Ibid., 312.

5. CONCLUSION

What Would Augustine Think?

This study has reviewed the application of four distinct *jus post bellum* activities across three large-scale conflicts. The success of these activities has ranged from highly effective, as in the case of the Marshall Plan, to highly ineffective, as noted during the discussion of recent efforts in Afghanistan. However, the question posed by this study remains. Given that just war theorists have been retroactively attributing *jus post bellum* to Augustine, would Augustine approve of *jus post bellum* as it is practiced today? The answer, of course, is highly nuanced. Augustine may have approved of the post-Second World War *jus post bellum* activities, but he would not have approved of *jus post bellum* actions in recent years, even those that were deemed successful. The reason is that the very criteria that are used in the initial decision to enter a conflict have an effect on the success or failure of activities that take place afterward.

To begin with the Second World War, given that Japan attacked America first and that Germany declared war on America soon afterward, it would be fair to say that America's entry into the war was in accordance with Augustinian principles because, essentially, it was a war of self-defense. As Mattox writes, "Augustine clearly exonerates the Romans for fighting for the defense of the *patria*."[359] It may be possible to argue otherwise based on actions that America took before December 1941, such as trade embargoes and arms deliveries to Allied nations, but Japan was the first to declare the commencement of hostilities.

Many of the aspects of *jus post bellum* discussed in this work as applied to the Second World War were planned well before hostilities had ceased. This is a vitally important point when considering later *jus post bellum* activities. Humanitarian assistance began at least two years before the war ended and was to take place in those areas that had been completely pacified. The

---

[359] Mattox, *Augustine*, 47.

decision to hold leaders accountable was also made well in advance of the cessation of hostilities. These efforts were not perfect, but they worked well enough. The Tokyo and Nuremberg trials were, perhaps, the first of their kind and serve as examples that last until the present day. To use another example, humanitarian assistance efforts in Japan prevented certain starvation with the onset of winter in 1945.

It is true that some of the post-war measures, such as dividing Germany and forming a defensive alliance against the Soviet Union were practical measures that intersected with *jus post bellum* political and economic reconstruction efforts. However, Augustine would have deeply understood that postwar measures, though designed to create a lasting peace, include pragmatic, worldly calculations. Even in his own day, Roman emperors had to deal with rival foreign powers, barbarian incursions, and internal revolts.

After the Second World War, though, it would be fair to say that *jus post bellum* principles became detached from their supposedly Augustinian roots. In Iraq and in Afghanistan, the wars themselves could not have been considered justified under Augustinian principles. In Afghanistan, the decision to commence hostilities was not taken after careful deliberation and as a last resort. Planners failed to fully comprehend the tribal politics in the region and, in fact, began hostilities only three months after 9/11. It may be possible to make a better case for the invasion of Iraq given that Saddam would have indeed fit the description of an unjust, oppressive ruler. However, it would be difficult to assert that the invasion ultimately produced a relatively lower level of human suffering. Here, too, a failure to understand political and social dynamics caused problems during the period of occupation. These *jus ad bellum* issues clouded the *jus post bellum* efforts that came afterward.

In both Iraq and in Afghanistan, guerrilla warfare continued well after formal hostilities had ceased. In Iraq, Coalition forces were forced to contend with militias that were fueled by sectarian divides, foreign influence, the rapid demobilization of the Iraqi army, and the removal of Ba'ath Party members from their places of employment. In Afghanistan, the semi-independent tribes began to see NATO forces as occupiers who were supporting an increasingly corrupt government in Kabul.

The result was that, while some aspects of *jus post bellum* in these nations appeared to be successful, long-term security concerns severely hampered attempts to create a lasting sense of stability. In Iraq, it became exceedingly difficult for humanitarian agencies to provide direct aid to the populace because, while Coalition forces were numerous, they could not provide universal protection. For similar reasons, Iraq's infrastructure was difficult to rebuild because, it too, was constantly under threat. In Afghanistan, the Provincial Reconstruction Teams were able to make some progress in combining security elements with humanitarian and economic relief, but they could only provide a limited level of support.

Jus post bellum efforts to create political reforms fell far short of similar efforts in Europe and in Japan. While it may be true that some of this disparity in results can be attributed to cultural differences as well as the West's inability to account for them, the problem was much more profound. As discussed, the decision to have a traditional council of elders in Afghanistan select their government appeared successful, at first. It was the need to continually support this government, due to security concerns, after it had lost the respect of the populace that became problematic. In Iraq, at least, the new government continues to stand, though its long-term viability remains to be seen.

This is not to say that all efforts failed or that just war theory has not taken away valuable lessons from these experiences. In Iraq, the new government's decision to prosecute Saddam Hussein could have been considered a step forward as this was, possibly, the first time that such a leader was held accountable by his own people. Another possible lesson learned would be that a reformed government does not necessarily need to be a Western-style democracy so long as it serves the needs of the people. Augustine would have approved of this idea given that, within his own context, Rome was under authoritarian rule. The use of the *Loya Jirga* in Afghanistan appeared successful for a time and could serve as an example for future considerations of *jus post bellum* political reform in that region.

Moral and Apologetic Implications

It is also important to recall that there was a theological context behind Augustine's thoughts on war. The earthly city could never equal the splendor of the City of God, but all Christians had a duty to create the conditions that would point others in the direction of the heavenly city to the greatest extent possible. For leaders, this meant establishing internal and external harmony, supporting the work of the Church, and partnering with its citizens in the pursuit of these goals. The rise of the all-volunteer military makes Augustine's thoughts on war especially relevant in a present-day context. Augustine saw Christians who served in the Roman military as citizen soldiers who committed violence on behalf of the state.[360] For the good of the souls of these soldiers, it was important that any violence be legal and permissible.

Though just war theory, as practiced today, is more of a secular construct, the key requirement to protect the moral well-being of military members remains. To put it simply, it is vital to the moral and spiritual well-being of those who commit violence on behalf of the state that they feel that their actions are both proper and justified. In his commentary on Augustine's *Reply to Faustus the Manichaean*, Iasiello writes, "Augustine was concerned that in the midst of the chaos of war people might lose their human focus. In expressing this concern, he displays the heart and sensitivity of a military chaplain."[361] Given that the reasons for going to war in Iraq and Afghanistan have been difficult to justify, it may also be difficult for returning military members to reconcile their own actions within these conflicts. Therefore, when considering future *jus post bellum* activities, the justification for the war and the planning for the post-war activities encompasses more than national success or failure. These factors have a direct impact on the moral state of the nation and its citizens.

This study also matters because, now that recent *jus post bellum* activities have failed, the opportunity may be available to reassert an Augustinian view of *jus post bellum*. In a present-day context, what principles would be in place? *Jus ad bellum* and *jus in bello* could be a guide. As discussed in the historical overview of just war development, the sacred underpinnings of these concepts slowly gave way to more secular moral concerns, but underlying tenets such as just cause, and proper treatment of civilians remained. For *jus post bellum*, the principal tenet that would remain would be the necessity of ensuring that, when a war ends, the final result is a lasting peace.

For Augustine, a lasting peace may have certainly provided a safe environment for the Church to perform its necessary duties and for the populace to focus on spiritual concerns. This dynamic may have changed since his time, but it remains true today that a relatively stable environment has a positive effect on ministries taking place in a given country and that people have a greater ability to focus on spiritual matters when they do not need to worry about immediate material and safety needs.

---

[360] Augustine, *The City of God: Books I-VII*, 53.
[361] Iasiello, "*Jus Post Bellum*," 50.

Conclusion

The four pillars of *jus post bellum*, as defined in this study, have a proven ability to establish a lasting peace provided that they are performed correctly. Rebuilding an economy, providing humanitarian aid, holding unjust leaders accountable, and establishing a respected government worked so well in Europe and in Japan that those nations remain prosperous to the present day. While the same level of success in Iraq and in Afghanistan may never have been possible, there were clear indicators that an acceptable level of success was obtainable provided that the wars were carried out for the correct reasons and after careful consideration, that cultural concerns were fully internalized and respected, and that the wars were fully concluded before *jus post bellum* efforts began.

Final Thoughts

When this work was first proposed as a dissertation, it was under the assumption that its research questions filled a relatively small void in current scholarship. With the conclusion of American involvement in Afghanistan, it may be possible that a much larger field of study is being opened. What does the rapid collapse of the Kabul government tell us? Did two decades of Western-style education of Afghan women have a positive lasting impact? Questions such as these, and more, may be of interest to scholars in the future.

Of greater relevance to this study, and the works that could come after it, are the questions surrounding the future of *jus post bellum*. Now that the most recent *jus post bellum* effort has apparently failed to bear fruit, though the long-term implications are yet to be seen, deep questions remain regarding the future of post-war involvement as well as its Augustinian context. At the end of the Second World War, it was apparent that, while *jus post bellum* activities were imperfectly applied and pragmatic considerations often took priority, they were lasting, effective, and could be traced reasonably well to the Augustinian roots that scholars had assigned to them. Now that more recent *jus post bellum* efforts have produced very different outcomes, how should America practice its post-war activities in the future? This is as much a moral and theological question as it is a military and legal question.

This is a moral question because, within the context of an all-volunteer military, it is important that those who fight do so with the belief that they are committing morally just actions and that they are risking their well-being for the greater good. While it is true that much of this issue lies with the initial decision to go to war, the actions taken afterward have a profound effect on all parties involved. Augustine spoke at length about this issue. Furthermore, for the good of the state and its populace, there needs to be a sense that the nation can be proud of its post-war conduct. As this study notes, there was this sense after the conclusion of the Marshall Plan, but not after the conclusion of more recent conflicts.

This is a theological question because, even though just war theory has moved from a sacred to a secular context, the idea of providing practical aid and creating a level of security that would be conducive to focus on spiritual matters, as well as the safe operation of the Church, is important. This shift from the sacred to the secular occurred over time both from a just war perspective and a cultural perspective. In terms of just war theory, the shift occurred slowly as *jus ad bellum* and *jus in bello* developed over time. A concurrent cultural shift has also taken place to the point where, as Charles Taylor notes, "You can engage fully in politics without ever encountering God."[362] Still, the Church continues to perform its work in what could be considered a post-Christian world, and *jus post bellum*, applied properly, could continue to provide indirect support to the Church's endeavors.

Furthermore, since the future of *jus post bellum*, which has been traced to Augustine, is an unknown, it is possible that the opportunity exists for the Church to attempt to influence *jus post bellum* in an effort to bring it back to what could be considered an Augustinian just war theory. Future conflict is inevitable, and, in the interim, the Church has the tools that it needs to voice its concerns on a national stage and in the development of public policy. This period of relative peace may be, perhaps, a unique opportunity for the Church to have a direct, positive influence on the conclusion of future conflicts.

In terms of specific areas for further research, two proposals immediately come to mind. First, it remains to be seen whether or not the two decades of Western-style education have had a lasting impact on the women of Afghanistan and on the culture as a whole. Now that the Taliban has regained control of the country and has taken Afghan women out of schools, it would be interesting to examine what impact they have on their culture over the next several years in comparison to their involvement prior to 2001. It is likely that the data for this study does not yet exist, but the necessary data may be available relatively soon. This proposed study could have an impact on the reformation of an Augustinian *jus post bellum* concept

---

[362] Charles Taylor, *A Secular Age* (Cambridge, Massachusetts and London, England: The Belknap Press of Harvard University Press, 2007), 1.

if education is found to be a stabilizing force alongside other *jus post bellum* pillars presented in this study.

Second, this work has asserted that the Church should involve itself in the reformation of an Augustinian *jus post bellum* construct, but it has not stated, specifically, what form that involvement would take. One possible avenue for reform could be to leverage the influence of the military chaplains of each service. Iasiello, who has been quoted extensively in this study, appeared to have been such an advocate and, indeed, he penned his thoughts in a peer- reviewed journal read widely by senior military leaders. Therefore, an idea for further research may be to ascertain the standpoints of other senior chaplains on this matter as part of a much larger effort to encourage advocacy for a more Augustinian *jus post bellum* effort in the future.

BIBLIOGRAPHY

Allawi, Ali A. *The Occupation of Iraq: Winning the War, Losing the Peace*. New Haven, CT: Yale University Press, 2007.

Armstrong-Reid, Susan, and David Murray. *Armies of Peace: Canada and the UNRRA Years*. Toronto, Canada: University of Toronto Press, 2008.

Augustine of Hippo. *A Treatise Concerning the Correction of the Donatists*. In *A Select Library of the Nicene and Post-Nicene Fathers of the Christian Church*, Volume IV, *St. Augustine: The Writings Against the Manichaeans and the Donatists*. Translated by J.R. King. Edited by Philip Schaff. Buffalo, NY: The Christian Literature Company, 1887.

_____. *Letters of St. Augustine*. In *A Select Library of the Nicene and Post-Nicene Fathers of the Christian Church*. Volume I, *The Confessions and Letters of St. Augustine with a Sketch of His Life and Work*. Translated by J.G. Cunningham. Edited by Philip Schaff. Buffalo, NY: The Christian Literature Company, 1886.

_____. *Reply to Faustus the Manichaean*. In *A Select Library of the Nicene and Post-Nicene Fathers of the Christian Church*. Volume IV, *St. Augustine: The Writings Against the Manichaeans and Against the Donatists*. Translated by Richard Stothert. Edited by Philip Schaff. Buffalo, NY: The Christian Literature Company, 1887.

_____. *Selected Writings*. Translated by Mary T. Clark. Mahwah, NJ: Paulist Press, 1984.

_____. *The City of God: Books I-VII*. Translated by Demetrius B. Zema and Gerald G. Walsh. Washington, D.C.: The Catholic University of America Press, 2008.

_____. *The City of God: Books VIII-XVI*. Translated by Gerald G. Walsh and Grace Monahan. Washington, D.C.: The Catholic University of America Press, 2008.

_____. *The City of God: Books XVII-XXII*. Translated by Gerald G. Walsh and Daniel J. Honan. Washington, D.C.: The Catholic University of America Press, 2008.

Bailey, Beth. *America's Army: Making the All-Volunteer Force*. Cambridge, MA: Harvard University Press, 2009.

Barfield, Thomas. *Afghanistan: A Cultural and Political History*. Princeton, NJ: Princeton University Press, 2010.

Barnes, Timothy. *Constantine: Dynasty, Religion and Power in the Later Roman Empire*. West Sussex, UK: Wiley Blackwell, 2014.

Bass, Gary J. "Jus Post Bellum." *Philosophy and Public Affairs* 32, no. 4 (2004): 384-412.

Bellamy, Alex J. "The Responsibilities of Victory: *Jus Post Bellum* and the Just War." *Review of International Studies* 34 (2008): 601-625.

Bothe, Michael, K.J. Partsch, and W.A. Solf. *New Rules for Victims of Armed Conflicts: Commentary on the Two 1977 Protocols Additional to the Geneva Conventions of 1949*. Boston, MA: Brill, 2013.

Brown, Peter. *Augustine of Hippo*. Los Angeles, CA: University of California Press, 1970.

Carruthers, Bob, ed. *The Gestapo on Trial: Evidence from Nuremberg*. South Yorkshire, UK: Pen and Sword Books, 2014.

Carvin, Stephanie. "Francis Lieber (1798-1872)." In *Just War Thinkers: From Cicero to the 21st Century*, edited by Daniel R. Brunstetter and Cian O'Driscoll, 180-192. London, UK: Routledge, 2017.

Cavanna, Thomas P. *Hubris, Self-Interest, and America's Failed War in Afghanistan: The Self-Sustaining Overreach*. London, UK: Lexington Books, 2015.

Cole, Darrell. "The First and Second Gulf Wars." In *America and the Just War Tradition: A History of U.S. Conflicts*, edited by Mark David Hall and J. Darryl Charles, 251-270. Notre Dame, IN: University of Notre Dame Press, 2019.

Cox, Rory. "Gratian (Circa 12th Century)." In *Just War Thinkers: From Cicero to the 21st Century*, edited by Daniel R. Brunstetter and Cian O'Driscoll, 34-49. London, UK: Routledge, 2017.

Coyne, Christopher J., and Adam Pelillo. "Economic Reconstruction Amidst Conflict: Insights from Afghanistan and Iraq." *Defense and Peace Economics* 22, no. 6 (2011): 627-643.

Cryer, Robert, and Neil Boister. *The Tokyo International Military Tribunal – A Reappraisal*. Oxford, UK: Oxford University Press, 2008.

Demont-Biaggi, Florian. "Causation, Luck, and Restraint in War." In *Jus Post Bellum: Restraint, Stabilisation, and Peace*, edited by Patrick Mileham, 33-47. Boston, MA: Brill, 2020.

Dromi, Shai. *Above the Fray: The Red Cross and the Making of the Humanitarian NGO Sector*. Chicago, IL: University of Chicago Press, 2020.

Dyson, R.W., trans. *Aquinas: Political Writings*. Cambridge, UK: Cambridge University Press, 2002.

Ehrman, Bart D. *The Triumph of Christianity: How a Forbidden Religion Swept the World*. New York, NY: Simon & Shuster, 2018.

Eichbauer, Melodie H. "The Bishop with Two Hats: Reconciling Episcopal and Military Obligations in Causa 23 of Gratian's *Decretum*" in *Civilians and Warfare in World History*, edited by Nicola Foote and Nadya Williams, 120-139. New York, NY: Routledge, 2018.

Eusebius of Caesarea. *The Life of the Blessed Emperor Constantine*. In *Eusebius: Church History, Life of Constantine the Great, and Oration in Praise of Constantine*, edited by Philip Schaff and Henry Wace, translated by Ernest Richardson, 408-559. New York, NY: The Christian Literature Company, 1890.

Evans, G. R. *The Roots of the Reformation: Tradition, Emergence, and Rupture*. 2nd ed. Downers Grove, IL: IVP Academic, 2012.

Ferguson, Everett. *Church History Volume One: From Christ to Pre-Reformation*. Grand Rapids, MI: Zondervan, 2005.

Finn, Richard B. *Winners in Peace: MacArthur, Yoshida, and Postwar Japan*. Los Angeles: University of California Press, 1992.

Forsythe, David P., and Barbara Ann J. Rieffer-Flanagan. *The International Committee of the Red Cross: A Neutral State Actor*. New York, NY: Routledge, 2007.

Fotion, Nicholas. *War and Ethics: A New Just War Theory*. New York, NY: Continuum, 2007.

Futamara, Madoka. "Individual and Collective Guilt: Post-War Japan and the Tokyo War Crimes Tribunal." *European Review* 14, no. 4 (2006): 471-483.

Grotius, Hugo. *The Rights of War and Peace*. Edited by Knud Haakonssen. Indianapolis, IN: Liberty Fund, 2005.

Hanish, Shak. "Christians, Yazidis, and Mandaeans in Iraq: A Survival Issue." *Domes* 18, no. 1 (Spring 2009): 3-18.

Harmless, William, ed. *Augustine in His Own Words*. Washington, D.C.: Catholic University of America Press, 2010.

Hause, Jeffrey, ed. *Aquinas's Summa Theologiae: A Critical Guide*. Cambridge, UK: Cambridge University Press, 2018.

Heather, Peter J. *The Fall of the Roman Empire: A New History of Rome and the Barbarians*. New York, NY: Oxford University Press, 2006.

Heller, Kevin Jon. *The Nuremberg Military Tribunals and the Origins of International Criminal Law*. New York, NY: Oxford University Press, 2011.

Henderson, David E., and Frank Kirkpatrick. *Constantine and the Council of Nicaea: Defining Orthodoxy and Heresy in Christianity, 325 CE*. Chapel Hill, NC: University of North Carolina Press, 2016.

Hogan, Michael J. *The Marshall Plan: America, Britain, and the Reconstruction of Western Europe, 1947-1952*. Cambridge, UK: Cambridge University Press, 1987.

Holm, Michael. *The Marshall Plan: A New Deal for Europe*. New York, NY: Routledge, 2017.

Iasiello, Louis V. "*Jus Post Bellum*." *Naval War College Review* 57, no. 3 (2004): 33-52.

Ikenberry, G. John. *After Victory: Institutions, Strategic Restraint, and the Rebuilding of Order after Major Wars*. Princeton, NJ: Princeton University Press, 2001.

Johnson, James Turner. "Aquinas and Luther on War and Peace: Sovereign Authority and the Use of Armed Force." *The Journal of Religious Ethics* 31, no. 1 (Spring 2003): 3-20.

Junker, Detlef, Philipp Gassert, Wilfried Mausbach, and David B. Morris. *The United States and Germany in the Era of the Cold War, 1945-1990*. Cambridge, UK: Cambridge University Press, 2004.

Kaufman, Peter Iver. "Augustine's Punishments." *Harvard Theological Review* 109, no. 4 (October 2016): 550-566.

Klein, Albert W. "Attaining the Post Conflict Peace Using the *Jus Post Bellum* Concept." *Religions* 11, no. 4 (2020): 173-194.

Lang, Jr., Anthony F. "Hugo Grotius (1583-1645)." In *Just War Thinkers: From Cicero to the 21st Century*, edited by Daniel R. Brunstetter and Cian O'Driscoll, 128-143. London, UK: Routledge, 2017.

Large, Stephen. *Emperor Hirohito and Showa Japan: A Political Biography*. New York, NY: Routledge, 1992.

Leavitt, William M. "General Douglas MacArthur: Supreme Public Administrator of Post-World War II Japan." *Public Administration Review* 75, no. 2 (March/April 2015): 315-324.

Lee, Peter. "Selective Memory: Augustine and Contemporary Just War Discourse." *Scottish Journal of Theology* 65, no. 3 (July 2012): 309-322.

Levering, Matthew. *The Theology of Augustine: An Introductory Guide to His Most Important Works*. Grand Rapids, MI: Baker Academic, 2013.

Lischer, Sarah Kenyon. "Military Intervention and the Humanitarian 'Force Multiplier'." *Global Governance* 13 (2007): 99-118.

Lorenzo, David J. *War and American Foreign Policy: Justifications of Major Military Actions in the US*. Taipei: Palgrave Macmillan, 2021.

Luther, Martin. "Temporal Authority: To What Extent It Should Be Obeyed" in *Martin Luther's Basic Theological Writings*. 3rd ed. edited by Timothy F. Lull and William R. Russell, 428-455. Minneapolis, MN: Fortress Press, 2012.

Lykogiannis, Athanasios. *Britain and the Greek Economic Crisis, 1944-1947: From Liberation to the Truman Doctrine*. Columbia, MO: University of Missouri Press, 2002.

MacCulloch, Diarmaid. *The Reformation: A History*. New York: NY: Penguin Books, 2004.

Maley, William. *Transition in Afghanistan: Hope, Despair, and the Limits of Statebuilding*. London, UK: Routledge, 2018.

Markham, J. David. *The Road to St. Helena: Napoleon after Waterloo*. South Yorkshire, UK: Pen and Sword, 2008.

Marsden, Peter. *Afghanistan: Aid, Armies, and Empires*. London, UK: I.B. Tauris, 2009.

Mattox, John Mark. *Saint Augustine and the Theory of Just War*. New York, NY: Continuum Books, 2006.

Mayers, David. *America and the Postwar World: Remaking International Society, 1945-1956*. London, UK: Routledge, 2018.

McLeary, Rachel M. *Global Compassion: Private Voluntary Organizations and U.S. Foreign Policy Since 1939*. New York, NY: Oxford University Press, 2009.

Metaxas, Eric. *Martin Luther: The Man Who Rediscovered God and Changed the World*. New York, NY: Penguin Books, 2018.

Moore, Aaron Stephen. *Constructing East Asia: Technology, Ideology, and Empire in Japan's Wartime Era, 1931-1945*. Stanford, CA: Stanford University Press, 2013.

Neiberg, Michael S. *The Treaty of Versailles: A Very Short Introduction*. Oxford, UK: Oxford University Press, 2018.

Newton, Michael. "Community-Based Accountability in Afghanistan: Recommendations to Balance the Interests of Justice." In *Jus Post Bellum and Transitional Justice*, edited by Larry May and Elizabeth Edenberg, 74-112. Cambridge, UK: Cambridge University Press, 2013.

Novikova, Liudmila. *An Anti-Bolshevik Alternative: The White Movement and the Civil War in the Russian North*. Translated by Seth Bernstein. Madison, WI: University of Wisconsin Press, 2018.

O'Driscoll, Cian. "No Substitute for Victory? Why Just War Theorists Can't Win." *European Journal of International Relations* 26, no. 1 (July 2019): 187-208.

Orend, Brian. "Immanuel Kant (1724-1804)." In *Just War Thinkers: From Cicero to the 21st Century*, edited by Daniel R. Brunstetter and Cian O'Driscoll, 168-179. London, UK: Routledge, 2017.

———. "*Jus Post Bellum*: The Perspective of a Just-War Theorist." *Leiden Journal of International Law* 20 (2007): 571-591.

Patterson, Eric D. *Ending Wars Well: Order, Justice, and Conciliation in Contemporary Post-Conflict*. New Haven, CT: Yale University Press, 2012.

Pattison, James. "Jus Post Bellum and the Responsibility to Rebuild." *British Journal of Political Science* 45, no. 3 (2015): 635-661.

Pendas, Devin O. *Democracy, Nazi Trials, and Transitional Justice in Germany, 1945-1950*. Cambridge, UK: Cambridge University Press, 2020.

Pike, Francis. *Hirohito's War: The Pacific War, 1941-1945*. London, UK: Bloomsbury Publishing, 2016.

Reichberg, Gregory M. "Thomas Aquinas on Military Prudence." *Journal of Military Ethics* 9, no. 3 (2010): 262-275.

———. *Thomas Aquinas on War and Peace*. New York, NY: Cambridge University Press, 2017.

Rich, John, and Graham Shipley, eds. *War and Society in the Roman World*. London, UK: Routledge, 1993.

Roberts, Adam. "Foundational Myths in the Laws of War: The 1863 Lieber Code, and the 1864 'Geneva Convention'." *Melbourne Journal of International Law* 20, no. 1 (July 2019): 158-196.

Sassoon, Joseph. "Iraq's Political Economy Post 2003: From Transition to Corruption," *International Journal of Contemporary Iraqi Studies* 10, no. 1-2 (March 2016),17-33.

Scalise, Charles J. "Exegetical Warrants for Religious Persecution: Augustine vs. the Donatists." *Review and Expositor* 93 (1996): 497-505.

Schaller, Michael. *The American Occupation of Japan: The Origin of the Cold War in Asia*. New York, NY: Oxford University Press, 1987.

Slim, Hugo. *Humanitarian Ethics: A Guide to the Morality of Aid in War and Disaster*. Oxford, UK: Oxford University Press, 2015.

Smith, Charles Anthony. *The Rise and Fall of War Crimes Trials: From Charles I to Bush II*. Cambridge, UK: Cambridge University Press, 2012.

Smither, Edward L. *Augustine as Mentor: A Model for Preparing Spiritual Leaders*. Nashville, TN: B&H Academic, 2009.

Stahn, Carsten. "*Jus Post Bellum*: Mapping the Discipline(s)." *American University International Law Review* 23, no. 2 (2007): 311-347.

Stahn, Carsten, Jennifer S. Easterday, and Jens Iverson. *Jus Post Bellum: Mapping the Normative Foundations*. Oxford, UK: Oxford University Press, 2014.

Steeves, Rouven. "The War on Terror and Afghanistan." In *America and the Just War Tradition: A History of U.S. Conflicts*, edited by Mark David Hall and J. Darryl Charles, 271-299. Notre Dame, IN: University of Notre Dame Press, 2019.

Taye, Safiullah. "Afghanistan's Political Settlement Puzzle: The Impact of the Breakdown of Afghan Political Parties to an Elite Polity System (2001-2021)." *Middle East Critique* 30, no. 4 (October 2021), 333-352.

Taylor, Charles. *A Secular Age*. Cambridge, Massachusetts, and London, England: The Belknap Press of Harvard University Press, 2007.

Totami, Yuma. *The Tokyo War Crimes Trial: The Pursuit of Justice in the Wake of WW II*. 1st ed. Boston, MA: Harvard University Asia Center, 2008).

Van Dam, Raymond. *The Roman Revolution of Constantine*. New York, NY: Cambridge University Press, 2008.

Vorster, Nico. "Just War and Virtue: Revisiting Augustine and Thomas Aquinas." *South African Journal of Philosophy* 34, no. 1 (March 2015): 55-68.

Walzer, Michael. *Arguing about War*. New Haven, CT: Yale University Press, 2004.

Weiss, Daniel H. "Christians as Levites: Rethinking Early Christian Attitudes toward War and Bloodshed via Origen, Tertullian, and Augustine." *Harvard Theological Review* 112, no. 4 (October 2019): 491-505.

Wilken, Robert Louis. *The Christians as the Romans Saw Them*. 2nd ed. New Haven, CT: Yale University Press, 2003.

Williams, Howard. *Kant and the End of War: A Critique of Just War Theory*. New York, NY: Palgrave Macmillan, 2012.

Wynn, Philip. *Augustine on War and Military Service*. Minneapolis, MN: Augsburg Fortress Publishers, 2013.

Zwitter, Andrej, and Michael Hoelzl. "Augustine on War and Peace." *Peace Review* 26, no. 3 (August 2014): 317-324.

Index

Adam, 61, 102, 136, 140
Afghan, 71, 119–122, 131–132, 140
Afghanistan, 11–12, 14–17, 25–26, 71–72, 74–75, 77, 80, 102, 110, 113–123, 125, 127–132, 135–136, 138–140
Afghans, 75, 115, 117, 119–121
Africa, 10, 34–37
African, 20, 34, 141
Alaric, 35–36, 42
Albert, 15, 23, 137
Alex, 23, 135
Allawi, 74, 98, 100, 106, 109, 134
Allied, 63–64, 73, 87, 89, 91–93, 128
Amalek, 45–46
Amalekites, 45
American, 8, 11–12, 17, 25, 61, 63, 67, 69–72, 80, 85–90, 92–94, 96, 98, 108–109, 113, 115–118, 125, 131, 138, 140
Americans, 85, 93
Amorites, 46
Antiquity, 7, 12, 18, 24, 27
Aquinas, 12–13, 20–21, 55–59, 95, 136–137, 139, 141
Arab, 106
Arabian, 102
Arabic, 106
Arabs, 34, 53, 74, 98, 100, 103
Aristotle, 7, 95
Armstrong, 91, 134
Asia, 63, 66, 81, 83, 89–90, 94, 139–140
assassination, 62, 99–100
asylum, 115, 117
Athanasios, 69, 138
Augustine, 4, 7–14, 16, 18–25, 27–31, 33–62, 64, 67, 72, 75, 79–81, 87, 90, 94–97, 101, 107–108, 110, 112–113, 116, 123, 125–130, 132, 134–135, 137–138, 140–141
Augustinian, 1, 8–9, 11–15, 18, 25–26, 33, 43–44, 55–64, 67, 69, 72, 75, 78, 80, 84, 87, 90, 92, 94–97, 101, 105, 110, 112–113, 116, 118, 121, 123–125, 127–128, 130–133
authoritarian, 107–108, 110, 113, 116, 119, 125, 129
Axis, 90–91

Baghdad, 74, 106, 109
Bailey, 113, 135
Baker, 138
Balkans, 97
barbarian, 35–36, 40, 128
Barbarians, 35, 137
Barfield, 74, 117–119, 121, 135
Barnes, 28–29, 69–70, 135
Bass, 14, 24, 66, 68, 75, 99, 103–104, 135
Belknap, 132, 140
Bellamy, 23, 107, 110, 135
belligerent, 11, 19, 23, 46, 60, 63, 67, 70, 74–75, 87, 125
belligerents, 26, 65–67, 70, 82
bello, 8–10, 12, 18, 21–22, 43, 58, 61–62, 66, 110, 130, 132
Bellum, 1, 4, 8, 12, 14–16, 20, 23–24, 63–66, 68, 70–71, 75, 78–79, 92, 99, 103–104, 106, 110–111, 115, 123, 130, 135–137, 139–140
bellum, 8–18, 20, 23–26, 30, 43–44, 48–49, 51, 60–66, 68, 70, 72, 74–76, 78, 80, 83–95, 97–98, 100–103, 105, 107–133
Bloomsbury, 64, 139
Bolshevik, 73, 139
Bonaparte, 65
Boniface, 37, 49
Boston, 64, 66, 77, 135–136, 140
Bothe, 77, 135
Bourbon, 73
Brill, 64, 77, 135–136
British, 70, 89, 139
Brunstetter, 12, 56, 59–61, 135, 138–139
Buffalo, 45, 48, 52, 134
Bush, 17, 115, 140

Caesar, 31
Caesarea, 28, 136
Cambridge, 34, 57, 66, 73, 84, 113, 115, 132, 135–137, 139–140
Canaanites, 46
Carruthers, 81, 135
Carsten, 8, 15, 63, 140
Carthage, 34–37, 42
Carvin, 61, 135
Catholic, 10, 34, 36, 44, 52, 57–58, 134–135, 137
Causa, 56, 136
Causation, 64, 136
Cavanna, 113–115, 135

Chapel, 29, 137
chaplain, 130
chaplains, 77, 133
Chicago, 77, 136
China, 83–84
Chinese, 84
Christendom, 19
Christian, 7–11, 18–20, 24, 28–35, 37–43, 45, 48, 51–52, 54–61, 72, 78, 95, 97, 108, 112, 123, 126, 132, 134, 136, 141
Christianity, 7, 19, 27–34, 39–42, 54, 113, 136–137, 142
Christianization, 11, 39–40, 54
Christianize, 28
Christianized, 7
Christianizing, 30
Christians, 9–11, 27–34, 37–43, 45, 50, 53, 58, 60, 72, 112–113, 126, 130, 137, 141
Christopher, 102, 136
Cian, 12, 56, 59–61, 135, 138–139
Cicero, 12, 56, 59–61, 135, 138–139
Cincinnatus, 42
Columbia, 69, 138
combatants, 19, 22, 24, 43, 50, 57, 77, 116, 124
Communist, 69–70
Constantine, 11, 20, 27–32, 34, 37, 39–40, 42, 53, 135–137, 140
Constantinian, 29, 39
Constantinople, 35
Constantius, 30
Cryer, 83–84, 93, 136
Cuba, 115
Cunningham, 48, 134
Cyprian, 34

Daniel, 11–12, 44, 56, 59–61, 135, 138–139, 141
Danube, 36
Darius, 48
David, 17, 29, 43, 65, 71, 73, 76, 90–91, 98, 134–138, 140
Decretum, 56, 136
democracies, 74–75, 87
democracy, 73–74, 87, 89, 107–108, 120, 129
democratic, 64, 73, 75, 87, 89, 93–94, 96, 98, 114, 120
Diarmaid, 138
Diocletian, 29, 34, 38–39
doctrine, 9, 15, 43, 48, 50, 61

Donatist, 9, 33, 51–52
Donatists, 9–10, 33–34, 45, 51–53, 116–117, 134, 140
Driscoll, 12, 56, 59–61, 135, 138–139
Dromi, 77, 136
Dujail, 99–100
Dyson, 57, 136

ecclesiastical, 31, 33, 53, 57
economic, 13, 18, 23–25, 35, 49, 63–64, 69–72, 81, 84–88, 91–93, 96–97, 101–106, 109–111, 117–118, 128–129
Ehrman, 32, 136
Eichbauer, 56, 136
Empire, 7, 10–11, 18, 27–28, 30–31, 34–40, 44–45, 52, 54, 72, 83, 90, 116, 135, 137, 139
empire, 18, 27–28, 30, 33, 39, 42, 108
Empires, 121, 138
England, 132, 140
Enlightenment, 55, 60
Episcopal, 56, 136
ethical, 8, 22, 24–26, 38, 50, 117
ethics, 7, 13, 18, 25, 95–96
European, 14, 62, 69, 73, 83–85, 88, 136, 139
Eusebius, 28, 30, 136
execution, 66, 81, 91, 99, 101, 111
Exodus, 45–46

Faustus, 9–10, 45–46, 130, 134
Ferdinand, 62
Ferguson, 31, 33–34, 136
Finn, 84, 86, 136
Flanagan, 76, 136
Flensburg, 73
Florian, 64, 136
Forsythe, 76, 136
Fotion, 62, 136
Francis, 61, 64, 135, 139
Frank, 29, 137
Franks, 35
Fray, 77, 136

Gauls, 35
Geiseric, 36–37
German, 58, 63, 66, 81, 87–88, 93–94

Gestapo, 81, 135
Gibraltar, 37
Goths, 35, 42–43
Graham, 36, 140
Gratian, 55–58, 135–136
Gratiani, 56
Greece, 69–70, 84
Gregory, 13, 55, 57, 139
Grotius, 55, 59–60, 137–138
Grove, 55, 136
Guantanamo, 115, 125
Gulf, 71, 80, 98, 101–102, 135

Haakonssen, 59, 137
Hamid, 119
Hanish, 113, 137
Harvard, 66, 113, 132, 135, 137, 140–141
Hause, 137
Henderson, 29, 137
Hill, 29, 137
Hippo, 9–10, 36–37, 44–45, 48, 52, 134–135
Hirohito, 64, 73, 138–139
Hitler, 73, 81
Hittites, 46
Hivites, 46
Hoelzl, 11, 20, 141
Hogan, 84, 137
Holocaust, 93
Howard, 60, 141
Hugo, 59, 76, 137–138, 140
Huns, 36
Hussein, 67, 74, 98–100, 106, 111–113, 129

Iasiello, 14, 23, 62–63, 65, 78–79, 110–111, 130, 133, 137
Ikenberry, 88, 137
Immanuel, 60, 139
insurgencies, 105–106, 111, 113, 117, 124
insurgency, 72, 98–99, 104–105, 107, 110, 120, 122
insurgents, 102, 104–105
invasion, 10, 35, 37, 86, 98–102, 104, 106, 117, 123, 125, 128
Invictus, 31
Iran, 98–99, 103
Iraq, 12, 14, 16–17, 25, 67, 71, 74–75, 77, 80, 98–114, 117–118, 120–123,

128–131, 134, 136–137, 140
Iraqi, 71, 98–108, 111, 128, 140
Iraqis, 71, 75, 108
Islam, 98
Islamic, 105–106, 113, 115
Israel, 45, 47
Israelites, 45–47
Italy, 63, 84
Ithaca, 69

Japan, 14, 17, 25, 63–64, 69–75, 80–81, 83–84, 86–90, 92–94, 97, 104, 112, 117–118, 127–129, 131, 136, 138–140
Japanese, 17, 69–70, 83–84, 86–90
Jebusites, 46
Joseph, 99, 101, 140
Joshua, 43, 46
Junker, 73, 137
jus, 8–18, 22–26, 30, 43–44, 48–49, 51, 58, 60–66, 68, 70, 72, 74–76, 78, 80, 83–95, 97–98, 100–103, 105, 107–133
Justice, 16, 65–68, 111, 115, 139–140
justice, 13, 16–17, 21, 23–24, 26, 43, 51, 59, 65–69, 78, 82–84, 86, 92–93, 95–97, 99, 101, 114–116, 125

Kabul, 74, 114–115, 118–122, 129, 131
Kant, 55, 60–61, 139, 141
Karzai, 119–120, 122
Kaufman, 137
Kenyon, 108, 138
Kirkpatrick, 29, 137
Knud, 59, 137
Kurdish, 99, 103
Kurds, 74, 98, 100, 103
Kuwait, 98, 100, 109

Lactantius, 28
leader, 20, 31, 45–46, 48, 54, 61, 98, 100, 129
leaders, 19–20, 22–23, 27, 29, 33, 37–38, 50, 56, 58, 63, 66–68, 73, 81, 83, 85, 95, 119, 122, 126, 128, 130–131, 133
leadership, 12, 45, 61, 78, 106–107
leading, 72, 101, 116
Leavitt, 86–87, 89, 92–93, 138
Lee, 112, 125–126, 138
Leiden, 65, 139

Levites, 11, 141
Lexington, 113, 135
Licinius, 29
Lieber, 61, 135, 140
Lischer, 108–110, 138
Liudmila, 73, 139
Luther, 13–14, 55, 58–59, 137–139
Lutheran, 13
Lykogiannis, 69, 138

MacArthur, 64, 84, 86–87, 89, 92–93, 136, 138
MacCulloch, 138
Madison, 73, 139
Madoka, 83, 136
Mahwah, 134
Mandaeans, 113, 137
Manichaean, 9, 45, 130, 134
Manichaeans, 45, 52, 134
Marcellinus, 41, 49
Markham, 65, 73, 138
Marsden, 121–122, 138
Marshall, 16–17, 23, 64, 69, 84–85, 87–88, 91, 93–94, 97, 127, 132, 137
Martin, 58, 138–139
Massachusetts, 132, 140
Mattox, 7–8, 10, 22, 25, 27–30, 41, 64, 72, 95–96, 127, 138
Mauretania, 37
Mausbach, 73, 137
medieval, 12–13
Mediterranean, 32, 42
Melbourne, 61, 140
Mesopotamia, 38
Metaxas, 139
Milan, 29, 39
Mileham, 64, 136
Milvian, 28, 39
Monahan, 135
Moore, 83, 139
moral, 12, 14, 16, 34, 40–42, 47–48, 55–59, 66, 70, 77, 79, 94–96, 99, 116, 126, 130–132
morally, 44, 46, 67, 76, 78, 96, 132
Moses, 45–47
Mughal, 118
Murray, 91, 134

Napoleon, 65–66, 73, 95, 138
Napoleonic, 65
NATO, 88, 90, 114–115, 118, 120–125, 129
Naval, 23, 137, 142
Nazi, 66, 81, 87, 139
Nazis, 93
NGOs, 76, 109, 122
Nicaea, 29, 137
Nicene, 45, 48, 52, 134
Nuremberg, 17, 23, 66, 81–84, 92–93, 100–101, 115–116, 128, 135, 137

Occupation, 69–70, 74, 89, 98, 100, 106, 109, 134, 140
occupiers, 46, 115, 121, 129
Orend, 60, 65, 71, 106–107, 139
Origin, 89, 140
Osama, 115, 119, 125
Oxford, 35, 63, 76–77, 81, 83, 89, 136–137, 139–140

Pacific, 64, 83–84, 86, 92, 139
pagan, 11, 18, 27–28, 30, 32, 38–42, 46–47, 52
Pakistan, 114–115, 122, 125
Paris, 65, 85
Pashtun, 72, 115
Pashtuns, 114, 117
Pashtunwali, 115, 117
Patterson, 16, 65–69, 94–96, 99–100, 106–108, 111, 139
Pattison, 70–71, 78, 111, 139
Paul, 29, 35, 106
Perizzites, 46
Peter, 10, 35, 47, 112, 121–122, 135, 137–138
Postmodern, 18, 25, 27, 31
PostNicene, 48
Postwar, 69, 84, 90, 136, 138
postwar, 15, 63, 69, 84, 96, 116, 128

Reichberg, 13, 55, 57–58, 139
Reid, 91, 134
reparations, 21, 63, 67–68, 107
Revolution, 34, 73, 140
Rhine, 36
Rhineland, 62
Rieffer, 76, 136

Roman, 7, 10–11, 18, 25, 27–28, 31–42, 44–45, 47–48, 52, 54, 56, 72, 90, 113, 116, 126, 128, 130, 135, 137, 140
Romans, 27, 31–32, 37, 42, 47, 127, 141
Rome, 10, 13, 31, 35–36, 39–42, 45, 54, 72, 107, 129, 137
Rumsfeld, 99
Russia, 73, 76
Russian, 73, 139

sacred, 18, 33, 59, 80, 130, 132
Saddam, 67, 74, 98–103, 106–108, 111–113, 128–129
Safiullah, 119, 140
Scottish, 112, 138
Shia, 103
Shiite, 74, 98, 103, 113
Shiites, 101, 106
Showa, 73, 138
Stahn, 8, 15, 63–64, 140
Stalin, 99
Summa, 57, 137
Sunni, 74, 98, 103, 106, 113

Taliban, 71, 74, 113–116, 118–122, 124–125, 132
Theodosius, 28, 32
torture, 50, 61, 66, 99
tortured, 124
torturing, 36
Trajan, 41
Truman, 69–70, 138
Turkey, 69, 102

Ukraine, 97
UN, 76, 100, 109, 122
UNHCR, 76
UNRRA, 90–92, 97, 134

Valentinian, 30
Vandal, 35
Vandals, 10, 34, 36–37
Vietnam, 113
Visigoths, 35

Weimar, 73

Yazidis, 113, 137
Yemen, 68

ABOUT THE AUTHOR

Edward Herty is a retired Naval Officer who earned a Ph.D. in Theology and Apologetics from Liberty University, an MBA from the Naval Postgraduate School, and an undergraduate degree in History from the U.S. Naval Academy. In this work, he integrates all three disciplines into a discussion of the relationship between Christianity and just war theory. He and his wife currently live in Northern Virginia.

www.ingramcontent.com/pod-product-compliance
Lightning Source LLC
LaVergne TN
LVHW020931090426
835512LV00020B/3309